/3000

THE HARTLEY TRANSFORM

THE HARTLEY TRANSFORM

by

Ronald N. Bracewell
Professor of Electrical Engineering
Stanford University

OXFORD UNIVERSITY PRESS • NEW YORK
CLARENDON PRESS • OXFORD
1986

Oxford University Press
Oxford New York Toronto
Delhi Bombay Calcutta Madras Karachi
Petaling Jaya Singapore Hong Kong Tokyo
Nairobi Dar es Salaam Cape Town
Melbourne Auckland

and associated companies in
Beirut Berlin Ibadan Nicosia

Published by Oxford University Press, Inc., 200 Madison Avenue,
New York, New York 10016

Library of Congress Cataloging-in-Publication Data

Bracewell, Ronald Newbold, 1921-
The Hartley Transform.

(Oxford engineering science series ; 19)
Includes index.
1. Hartley transforms. I. Title. II. Series.
QA403.5.B73 1986 515.7'23 86-13595
ISBN 0-19-503969-6

Printing (last digit): 9 8 7 6 5 4 3 2 1
Printed in the United States of America
on acid-free paper

PREFACE

While working on the material for this book the author had a number of pleasant experiences. One was to discover that good looking mathematics can be set rather readily if one has the assistance of my colleague Professor Donald Knuth's typesetting language TEX. As a person who never learned to type I am astonished at having been able to type this whole book. What is more, I could do all the illustrations. Many of the illustrations were simply computed with a Hewlett-Packard HP85 desktop microcomputer and drawn by an HP plotter which was kindly given by the Hewlett-Packard Company. In many interesting conversations Professor Oscar Buneman freely shared his knowledge of computation gained from experience with numerical simulation of plasmas. Former Stanford colleague Professor F. W. Crawford, now Vice-Chancellor of Aston University, proposed this project to Oxford University Press who invited me to consider it last October, precipitating me into six months of hard but enjoyable work. My wife Helen generously supported my exertions.

The ideas expounded here are to some extent original, but are more or less implicit in publications going back over forty years. I am happy to have played a role in articulating a complex of practices into the elegant framework provided years ago by Hartley.

Stanford, California
March, 1985

R.N.B.

Postscript added in proof. I am indebted to Mr. Craig H. Barratt, who kindly helped by perusing the manuscript and ably detecting a dozen nonobvious errors.

CONTENTS

CHAPTER 1

INTRODUCTION

"Vous savez le latin, sans doute?" –
"Oui, mais faites comme si je ne le savais pas."
Molière, Le Bourgeois Gentilhomme

Harmonic analysis, to which the present book is a contribution, has a wonderful history, often thought to have started with Fourier (1768-1830), who is famous for the assertion that arbitrary functions could be represented as the sum of a trigonometric series. Much work in mathematics and physics stemmed from this controversial claim. Part of the nineteenth century development of mathematical analysis, which deals with such basic and now everyday notions as continuity and limits, grew from problems with Fourier series. As for physics, Lord Kelvin's opinion was that, "Fourier's theorem is not only one of the most beautiful results of modern analysis, but it may be said to furnish an indispensable instrument in the treatment of nearly every recondite question in modern physics."

One need only contemplate the formula

$$\tfrac{1}{2}x = \sin x - \tfrac{1}{2}\sin 2x + \tfrac{1}{3}\sin 3x + \ldots$$

given by Euler (1707-1783) to realize that, as with all evolution, there are antecedents. In fact, a case can be made that, as long ago as the second century, Claudius Ptolemaeus (a Greek of Alexandria) used the same basic idea. A trigonometric sum of the form

$$z(t) = a_1 e^{i\omega_1(t-t_1)} + a_2 e^{i\omega_2(t-t_2)} + a_3 e^{i\omega_3(t-t_3)} + \ldots$$

is, after all, only complex notation for a sum of rotating vectors that is also describable as the locus of a point on an epicycle rolling on a deferent circle whose center moves on a third circle. The idea of including an extra rotating vector with the amplitude, frequency and phase needed to account for better data thus goes back to ancient times and is exactly the method used today, as it was then, for calculating the position of a planet. Fourier's innovation was to insist on the generality of a sum of terms with integrally related angular velocities. As is well known, this proposition was stoutly resisted by the great French mathematicians of the day who quite properly demanded to be convinced. The area of contention narrowed to the question of necessary and sufficient conditions for the existence of Fourier's integral and it took more than a century for the discussion to culminate in generalized function analysis which tidied up everything.

Complex number theory greatly facilitated the handling of oscillating quantities. (To discuss alternating currents without the complex phasor and $e^{j\omega t}$ would today be unthinkable.) Consequently the theory of Fourier series took advantage of the theory of complex numbers and it came to seem natural that a periodic function $p(t)$ should be analyzed into complex components $a_n e^{i2\pi nt}$, where the coefficients a_n are now complex. Thus

$$p(t) = \sum_{n=-\infty}^{\infty} a_i e^{i2\pi nt}$$

and, in the limit, for a function $f(t)$ that is not periodic,

$$f(t) = \int_{-\infty}^{\infty} F(n) e^{i2\pi nt} \, dt.$$

At the same time we know very well that, without benefit of i at all,

$$p(t) = a_0 + \sum_{n=1}^{\infty} (a_n \cos 2\pi nt + b_n \sin 2\pi nt)$$

as Fourier tells us; so one understands that the use of complex exponentials is convenient rather than fundamental.

Hartley's 1942 formulation of a real integral transform made it possible to dispense with complex representation. Although his result was conspicuously published in the Proceedings of the Institute of Radio Engineers, and mentioned in textbooks such as S. Goldman, *Frequency Analysis, Modulation and Noise* (McGraw-Hill, 1948) and my own book, R.N. Bracewell, *The Fourier Transform and Its Applications* (McGraw-Hill, 1965), the momentum of standard practice prevailed. Possibly even earlier and equally unnoticed mentions will be recalled.

Hartley's cas function transform is presented in Chapter 2 and related to the Fourier transform. By analogy,

$$H(s) = \int_{-\infty}^{\infty} f(x) \operatorname{cas} 2\pi sx \, dx$$

is defined as the *Hartley Transform*. Theorems for the Hartley transform are presented succinctly for reference in Chapter 3.

By extension one may define the *Discrete Hartley Transform*

$$H(\nu) = N^{-1} \sum_{\tau=0}^{N-1} f(\tau) \operatorname{cas}(2\pi\nu/N)$$

whose properties are explored in Chapter 4. The factor N^{-1} ensures that $H(0)$ is equal to the mean value of $f(\tau)$, a feature in harmony with the convention that the Fourier coefficient a_0 be equal to the d.c. value of a periodic waveform.

Digital filtering, or discrete convolution, is a major application of transforms, and is the subject of Chapter 5.

The advent of images produced or processed by computer and displayed on a cathode ray screen has greatly extended the application of two-dimensional analysis and it is interesting to find that the idea of a real transform generalizes readily. Thus one can conceive of a transform *plane* having a reversible correspondence with another plane on which an object is defined. What distinguishes the Hartley transform plane, if we may so call it, apart from the fact that the transform values assigned to each point are real, is the absence of redundancy. In the Fourier transform plane, by contrast, the values are complex, and values at diametrically opposite points form complex conjugate pairs. Chapter 6 introduces two dimensions, listing many theorems in their extended form.

A practical consequence of attention to the real discrete transform is that it can be expressed as a matrix operation. A factorization of the matrix has been discovered which leads to a new factorization of the discrete Fourier transform matrix and thus to a new fast algorithm for spectral analysis and convolution, all in real terms. Chapter 7 presents the matrix view and discusses permutation.

The fast algorithm is dealt with in Chapter 8. Many interesting facets of fast spectral analysis are discussed. It is shown how practical programs break down into a number of separate parts, each of which contributes to the total running time and each having its own dependence on the size of the data sequence to be analyzed. The treatment goes beyond traditional complexity analysis by asymptotic counting of operations, making use of a timing diagram, or stripe diagram, to convey a size-dependent breakdown. Special attention is then given to each of the stripes with important conclusions about fast trigonometric functions, fast rotation and fast permutation. At the core of the new algorithm is the real transform that runs in half the time taken by the complex Fourier transform.

It is well known that an optical lens system can produce the Fourier transform of a coherent optical object and so it is natural to ask what bearing the recent developments may have for optics. A method for producing a two-dimensional real representation has been discovered and is described in Chapter 9. On the real transform plane the electric fields are in phase; consequently the information content is distributed entirely in the form of amplitude variation. The full ramifications of this independence from phase, a quantity that is not responded to by detectors of electromagnetic radiation below a certain wavelength, have yet to be worked out.

The work concludes with a collection of computer programs and an atlas of Hartley transforms to help launch those interested in applying or extending the various aspects of real transforms. Commercial use of some of the programs may be subject to licensing or other arrangements with the Board of Trustees of Stanford University, c/o Office of Technology Licensing, Stanford, California 94305.

Substantial sets of problems have been included and will be found to contain occasional items of information additional to what is in the text. However their main use is in connection with teaching. They reflect the author's experience that a short numerical exercise is an excellent aid to assimilating an abstract concept, just as a graphical or geometrical exercise enlarges one's appreciation of an analytic description, even though that description is sufficient.

It is easy to understand how the world of mathematical physics could be content with the traditional Fourier integral for more than one hundred years. Even the introduction of a radically new method of computation, the Fast Fourier Transform, at first had little impact, but in time came to revolutionize numerical practice as electronic computing took hold.

When the Fast Fourier Transform was brought into the limelight by Cooley and Tukey in 1965 it had an enthusiastic reception in the populous world of electrical signal analysis as the news spread via tutorial articles and special issues of journals. This ferment occasioned mild surprise in the world of numerical analysis, where related techniques were already known. Admirable sleuthing by M.T. Heideman, C.S. Burrus and D.H. Johnson (to appear in *Archive for History of the Exact Sciences*) has now traced the origins of the method back to a paper of C.F. Gauss (1777-1855) written in 1805, where he says, "Experience will teach the user that this method will greatly lessen the tedium of mechanical calculation."

A fascinating sidelight of the historical investigation is that Gauss's fast method for evaluating the sum of a Fourier series antedates the work on which Fourier's fame is based. We should hasten to add that Gauss's paper was not published until much later [*Collected Works, Vol. 3* (Göttingen: Royal Society of Sciences, 1876)], and we should remember that when Fourier introduced the idea of representing an arbitrary periodic function as a trigonometric series eminent mathematicians such as Lagrange resisted it.

It was not obvious that an even better algorithm might exist but the new factorization of the Fourier matrix describing the Discrete Fourier Transform is in fact faster than the Fast Fourier Transform. Efforts to capitalize on the fact that data can normally be assumed to be real had been only partly successful because a program that Fourier transforms real data cannot be used to retransform the (necessarily complex) coefficients from the transform domain back to the data domain. Therefore both a direct and a quite different reverse program must be stored and selected. Neither type of program possesses the reciprocity characteristic of the Fourier transform but pairs of such programs do achieve a time saving at the expense of storage space and selection decisions.

The fast algorithm based on the new factorization rather elegantly tidies the problem and commends itself to numerical analysts. It deals only in real numbers, which are the only kind of data we have in the world of experiment, and proceeds straight to the answers we need, which are also usually expressed in real terms, without the need to move into the domain of the complex. Since, as has been well known for a century and a half, Fourier's coefficients are complex, it is perhaps hard to accept that the complex numbers are a construct of the human mind rather than of nature. Of course we all accept that a power spectrum, such as an optical spectrum, is described by a real function of a real frequency variable f, but is not the strength $a_f^2 + b_f^2$ of that power spectrum to be computed as the squared modulus of the complex coefficient $a_f + ib_f$? Well, yes, that is one way. The other way, explained in this book, is to compute $[H(f)]^2 + [H(-f)]^2$, where $H(f)$ is the (real) Hartley transform. Phase can be found just as directly. In fact there is no purpose served by the complex Fourier transform that is not also served by the real Hartley transform.

CHAPTER 2

THE HARTLEY TRANSFORM

"Omne tulit punctum qui miscuit utile dulci."
(He won every point who combined the useful with the sweet.)

Horace, Ars Poetica

In his original paper in the *Proceedings of the Institute of Radio Engineers* in 1942, R.V.L. Hartley (1890-1970) laid emphasis on the strictly reciprocal character of the pair of integral formulae that he introduced and in the following section his notation will be followed so that the full symmetry can be appreciated. With the passage of time, however, some conventions have become customary in electrical engineering circles as a result of which the earlier formulation has acquired an archaic appearance. Consequently, after faithfully recording the historical form, we shall rewrite the relations in notation that is consistent with later practice; then the transform in its modern form will be defined as the *Hartley Transform*.

The original formulation

We start from a time dependent signal $V(t)$, which may be thought of as a voltage waveform of the kind that might be applied to the terminals of a telephone line. The waveform possesses a frequency spectrum that can be expressed through the Fourier transform. There are several schools of thought about notation for the Fourier transform and the one that is about to be given here is practically obsolete but forms a good temporary basis for the discussion of Hartley's original transform.

This version of the Fourier transform is written $S(\omega)$ and is defined by

$$S(\omega)=\frac{1}{\sqrt{2\pi}}\int_{-\infty}^{\infty}V(t)e^{-i\omega t}dt.$$

The quantity $S(\omega)$ is a complex function of the angular frequency variable ω, which itself assumes only real values. Thus for any given waveform $V(t)$ a Fourier transform, or complex spectrum, $S(\omega)$ can be calculated, which is unique for that waveform.

Delta functions

A short digression is desirable at this point. It is possible to think of functions for which the integral given above does not exist and a good deal of mathematical thought has been given to very intricate problems that can be imagined. There are two main categories to which such concerns apply. One category comprises functions that cannot arise in the physical world and in addition are of little if any physical interest; the other contains delta "functions" and their derivatives, which are of major physical interest even though they are entities of a sophisticated mathematical kind.

Delta "functions" are nevertheless a mainstay of physical thinking, which is where they originated. We pay no attention in the present work to the first category, as exemplified by functions that possess an infinite number of infinite discontinuities. As to delta "functions" we recognize that they are not functions but we adopt the machinery of ***generalized functions*** [M.J. Lighthill, *An Introduction to Fourier Analysis and Generalized Functions* (Cambridge University Press, 1958)] for dealing with $\delta(t)$ and $\delta'(t)$, which represent a unit impulse and its derivative respectively.

Null functions

Given $S(\omega)$, the question is, can the original waveform $V(t)$ be recovered? The answer is – mostly yes. One way of patching this imperfection would be to limit ourselves in advance to that class of functions for which the answer is yes. However it is rather useful to be aware of the exception because it is helpful in understanding a point about discontinuous functions that is often found perplexing and will be returned to later.

At this point it will suffice to explain the concept of a null function. The whole idea may sound faintly ridiculous but in fact it is a good idea to have in mind: a null function is one that integrates to nothing no matter how the limits of integration are chosen. Of course the function that is just plain zero is a null function. Another null function is $\delta^0(t)$ which is defined to be equal to 1 at $t = 0$ and zero elsewhere. Another is $\sum_i a_i \delta^0(t - t_i)$. These are the important ones. Null functions do not depend upon the cancellation of positives by negatives to achieve their null station; the infinite integral of the absolute value of a null function is also zero; that is, if $N(t)$ is a null function, then $\int_{-\infty}^{\infty} |N(t)|\, dt = 0$. While null functions are perfectly respectable mathematically, though of course the ones mentioned are discontinuous, they do not carry much weight in physics. If you apply a force $N(t)$ to a mechanical system there will be no effect.

Clearly $V(t) + N(t)$ will yield the same $S(\omega)$ as $V(t)$ alone. Therefore, when an attempt is made to invert the process, starting from $S(\omega)$ as given, it will not be possible to recover $V(t)$ in full detail in general, but only that part of it that remains when any null functions are omitted. If we omit them then we can say that $V(t)$ can be recovered from $S(\omega)$ by the inverse Fourier relation

$$V(t) = \frac{1}{\sqrt{2\pi}} \int_{-\infty}^{\infty} S(\omega) e^{i\omega t} d\omega.$$

The cas function integrals

Hartley introduced the different pair of formulae

$$\psi(\omega) = \frac{1}{\sqrt{2\pi}} \int_{-\infty}^{\infty} V(t) \operatorname{cas} \omega t \, dt$$

$$V(t) = \frac{1}{\sqrt{2\pi}} \int_{-\infty}^{\infty} \psi(\omega) \operatorname{cas} \omega t \, d\omega.$$

In these relations we follow the original paper in adopting the cas function, which is simply the sum of cos and sin, i.e.

$$\operatorname{cas} t \equiv \cos t + \sin t.$$

Thus there is not much apparent difference between the familiar Fourier transform integrals and the new pair of integrals, but in practice the difference is profound. For one thing, the transform $\psi(\omega)$ is real, not complex as will be the case in general with $S(\omega)$. Secondly, the inverse transformation calls for precisely the same integral operation as the direct transformation. Finally, $\psi(\omega)$ is not the familiar Fourier transform and we have to be prepared for unfamiliar properties and behavior. Much of the intuition that one may have built up about the Fourier transform or spectrum of a time-dependent waveform will not automatically apply to $\psi(\omega)$ and much of what follows is aimed at building up the experience that will make the real transform useful by pointing out differences and also the similarities.

The Hartley transform

In Hartley's definition of the transform $\psi(\omega)$ the factors $1/\sqrt{2\pi}$ were explicitly included in order to achieve a symmetrical appearance. If the factors are omitted, both integrals cannot be simultaneously correct. However, it has to be admitted that it is awkward to maintain a pair of such peculiar factors, especially in numerical work. Many authors have reacted to this in the case of the Fourier transform, where the same situation exists, by dealing with $\sqrt{2\pi}S(\omega)$ instead of $S(\omega)$. The result is that the factor disappears from the transform definition but a factor $1/2\pi$ appears in the inversion formula. Thus these authors deliberately sacrifice symmetry. It is true that a burden is placed on the memory, which is called upon to keep track of which formula has the 2π. One way of remembering is to note that the $1/2\pi$ comes with the integral that has the $d\omega$ and one knows that $\omega/2\pi$ is a combination that occurs naturally – in fact it is the same as frequency f. But having noted this fact, why not deal explicitly with frequency? This is the conclusion toward which consensus has been drifting for many years. The devotees of $1/\sqrt{2\pi}$ have practically died out, the $d\omega/2\pi$ people survive in various habitats, but general practice now heavily favors placing the 2π symmetrically in the exponent of both integrals. This happens automatically when frequency f is taken as the variable in place of angular frequency ω. Adopting this practice leads to

$$H(f) = \int_{-\infty}^{\infty} V(t) \operatorname{cas} 2\pi f t \, dt$$

$$V(t) = \int_{-\infty}^{\infty} H(f) \operatorname{cas} 2\pi f t \, dt.$$

Henceforth $H(f)$ will be taken as the Hartley transform of $V(t)$; and $V(t)$ will be the inverse Hartley transform of $H(f)$. Of course, the inverse transform is indistinguishable from the direct transform.

The Hartley transform as defined here is slightly different from Hartley's original transform $\psi(\omega)$. However, apart from historical reference it will hardly ever be necessary to mention $\psi(\omega)$ and should it be necessary a good recommendation is that $\psi(\omega)$ be called Hartley's transform. This fine distinction of terminology should be all that is needed. By introducing the new notation one gains consistency with majority usage as regards the Fourier transform, consistency that will later be carried over to the discrete transform.

For comparison, the Fourier transform relations written according to the same convention are

$$F(f) = \int_{-\infty}^{\infty} V(t) e^{-i2\pi ft}\, dt$$

$$V(t) = \int_{-\infty}^{\infty} F(f) e^{i2\pi ft}\, dt.$$

Even and odd parts

The relationship between the Fourier and Hartley transforms hinges upon symmetry considerations. To clarify this we split $H(f)$ into its even and odd parts, $E(f)$ and $O(f)$. The even part of a function is what we get when we reverse the function (changing t to $-t$), add the reversed function, or mirror image, to itself, and divide by two. Naturally the even part is its own mirror image, having the symmetrical property $E(-f) = E(f)$. The odd part is formed by subtracting the reversed function from the original and averaging; it has the antisymmetrical property $O(-f) = -O(f)$. Any function may be split uniquely into even and odd parts and from the even and odd parts, if they are given, the original function may be reconstituted uniquely. Among the interesting properties of the symmetrical and antisymmetrical parts is that the combined energies (or quadratic content) of the parts equals that of the whole.

To connect the transform $H(f)$ with the Fourier transform $F(f)$ of $V(t)$ it pays us to adopt the following definition.

Let $H(f) = E(f) + O(f)$, where $E(f)$ and $O(f)$ are the even and odd parts of $H(f)$ respectively. Then

$$E(f) = \frac{H(F) + H(-f)}{2} = \int_{-\infty}^{\infty} V(t) \cos 2\pi ft\, dt$$

and

$$O(f) = \frac{H(f) - H(-f)}{2} = \int_{-\infty}^{\infty} V(t) \sin 2\pi ft\, dt.$$

These two integrals are known as the Fourier cosine transform and the Fourier sine transform respectively and have been extensively tabulated [A. Erdélyi, *Tables of Integral Transforms, Vol. 1* (McGraw-Hill, 1954)].

Connecting relations

Given $H(f)$ we may form the sum $E(f) - iO(f)$ to obtain the Fourier transform $F(f)$:

$$E(f) - iO(f) = \int_{-\infty}^{\infty} V(t)(\cos 2\pi ft - i \sin 2\pi ft)\, dt$$

$$= \int_{-\infty}^{\infty} V(t) e^{-i2\pi ft}\, dt.$$

Thus we see that from $H(f)$ one readily extracts the Fourier transform of $V(t)$ by simple reflections and additions. The Fourier transform $F(f)$ has a real part which is the same as $E(f)$ and an imaginary part whose negative is $O(f)$:

$$F_{real}(f) = \text{Re}\, F(f) = E(f), \qquad F_{imag}(f) = \text{Im}\, F(f) = -O(f).$$

Conversely, given the Fourier transform $F(f)$, we may obtain $H(f)$ by noting that

$$H(f) = F_{real}(f) - F_{imag}(f):$$

from $F(f)$ one finds $H(f)$ as the sum of the real part and sign-reversed imaginary part of the Fourier transform.

Remembering that the imaginary part of a complex quantity is real, we verify that $F(f)$ is real, as it should be, given that the original waveform $V(t)$ was real. Had $V(t)$ itself not been real (in which case it could not have constituted values of a voltage waveform) then $F(f)$, and *a fortiori* $E(f)$ and $O(f)$, would not have been real. To summarize:

The Fourier transform is the even part of the Hartley transform minus i times the odd part; conversely, the Hartley transform is the real part of the Fourier transform minus the imaginary part.

Examples

Some particular transforms may be worked out for illustration.

Truncated exponential

As a first example take the function $\exp(-t)\mathrm{H}(t)$ which switches on at $t = 0$, rising to a value of unity, and then decaying exponentially. In this expression the Heaviside unit step function $\mathrm{H}(t)$ appears. It is defined by

$$\mathrm{H}(t) = \begin{cases} 0, & t < 0 \\ 1, & t > 0. \end{cases}$$

Note that no value is specified for $\mathbf{H}(0)$. The reason is this. Consider two functions $\mathbf{H}_a(t)$ and $\mathbf{H}_b(t)$ which are identical with $\mathbf{H}(t)$ where $t \neq 0$ but are specified at $t = 0$ and are specified differently. Let $\mathbf{H}_a(0) = a$ and $\mathbf{H}_b(0) = b$. Then the difference function $\mathbf{H}_a(t) - \mathbf{H}_b(t)$ is a null function. So, as far as integrals are concerned it makes no difference what finite value is assigned to $\mathrm{H}(0)$. As long as this is understood it seems more discreet to underspecify the integrand than to overspecify arbitrarily and trivially.

The integral to be evaluated is

$$\begin{aligned} H(f) &= \int_{-\infty}^{\infty} e^{-t}\mathrm{H}(t) \operatorname{cas} 2\pi ft \, dt \\ &= \int_{0}^{\infty} e^{-t} \operatorname{cas} 2\pi ft \, dt \\ &= \int_{0}^{\infty} e^{-t} \cos 2\pi ft \, dt + \int_{0}^{\infty} e^{-t} \sin 2\pi ft \, dt \\ &= \frac{1}{1 + 4\pi^2 f^2} + \frac{2\pi f}{1 + 4\pi^2 f^2}. \end{aligned}$$

By inspection it is clear that the even and odd parts are

$$E(f) = \frac{1}{1 + 4\pi^2 f^2}$$

and

$$O(f) = \frac{2\pi f}{1 + 4\pi^2 f^2}.$$

See Fig. 2.1 for a representation of the results. It may be noted that $H(f)$ possesses no symmetry, neither even nor odd. There is a minimum at $2\pi f = -1 - \sqrt{2}$, a maximum at $2\pi f = \sqrt{2} - 1$ and a zero crossing at $2\pi f = -1$. As $f \to \pm\infty$, $H(f)$ dies out as $\pm f^{-1}$.

For comparison with the Fourier transform we need the even and odd parts of the Hartley transform: the even part $E(f)$ is shown by a broken line and the odd part $O(f)$ by a dotted line. For the Fourier transform we have

$$\begin{aligned} F(f) &= E(f) - iO(f) \\ &= \frac{1}{1 + 4\pi^2 f^2} - i\frac{2\pi f}{1 + 4\pi^2 f^2}, \end{aligned}$$

a known result which can be confirmed separately. Each part possesses symmetry. As $f \to \pm\infty$ the real part dies out rapidly as $\pm f^{-2}$ whereas the

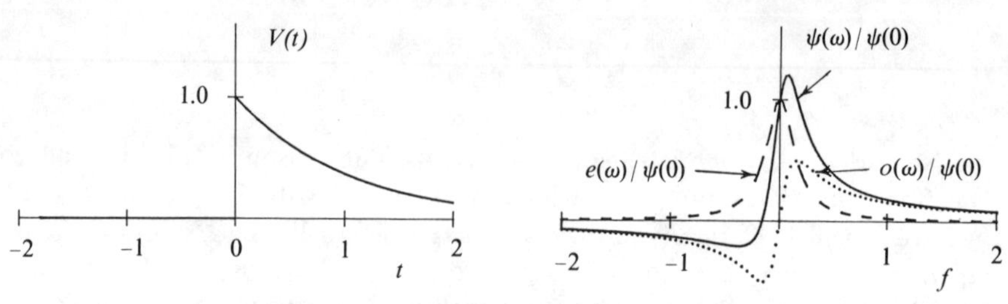

Fig. 2.1. The truncated exponential function $V(t) = e^{-t}\mathrm{H}(t)$ and its Hartley transform $H(f) = (1+2\pi f)/(1+4\pi^2 f^2)$. Note the symmetry of the even part $E(f)$ (broken line), the antisymmetry of the odd part $O(f)$ (dotted line), and the relatively rapid decay of the even part. The Fourier transform of $e^{-t}\mathrm{H}(t)$ is $E(f) - iO(f)$.

imaginary part dies out relatively slowly. For $f >> 1$ the imaginary part consequently dominates the spectrum. For example, at $f = 2$ the ratio of the imaginary part to the even part is $-12.6 : 1$.

Rectangular pulse

As a further example take $V(t) = \Pi(t - \frac{1}{2})$ as graphed on the left of Fig. 2.2. The factor $\Pi(t - \frac{1}{2})$ is a displaced unit rectangle function that switches on at $t = 0$. The standard unit rectangle function, which is frequently needed for gating out segments of waveforms, is defined by

$$\Pi(t) = \begin{cases} 0, & t < -\frac{1}{2} \\ 1, & -\frac{1}{2} < t < \frac{1}{2} \\ 0, & t > \frac{1}{2}. \end{cases}$$

For this example

$$\begin{aligned} H(f) &= \int_{-\infty}^{\infty} \Pi(t - \tfrac{1}{2}) \operatorname{cas} 2\pi f t \, dt \\ &= \int_0^1 \operatorname{cas} 2\pi f t \, dt \\ &= \frac{1}{2\pi f}[\sin 2\pi f t - \cos 2\pi f t]_0^1 \\ &= \frac{1}{2\pi f}(\sin 2\pi f - \cos 2\pi f + 1) \end{aligned}$$

as graphed in Fig. 2.2. Again one notes the lack of symmetry. For comparison, the Fourier transform is shown in Fig. 2.3. It is given by

$$F(f) = \frac{\sin 2\pi f}{2\pi f} - i\frac{1 - \cos 2\pi f}{2\pi f}.$$

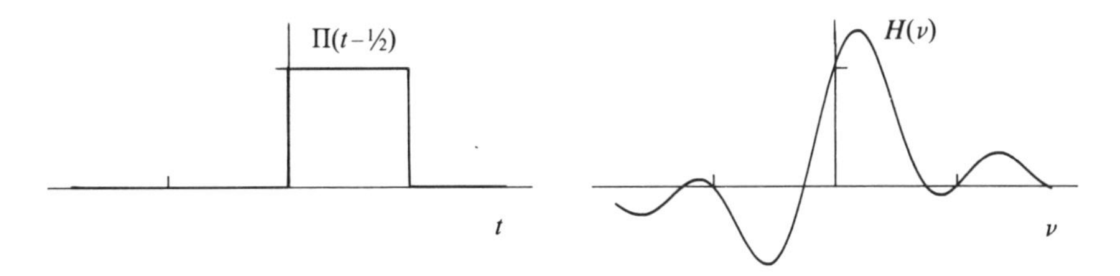

Fig. 2.2. A gate function $\Pi(t - \frac{1}{2})$ and its Hartley transform.

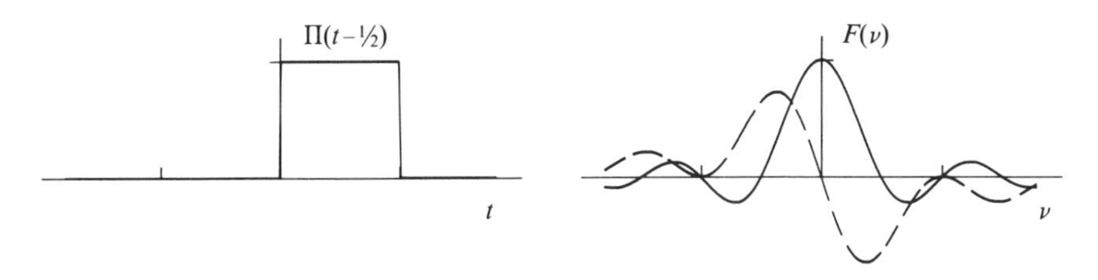

Fig. 2.3. The Fourier transform of the same gate function as in Fig. 2.2, the imaginary part shown as a broken line.

As usual, the real part is even and the imaginary part is odd. The two parts seem to die out at the same rate in this case, oscillating as they do so.

Impulse

As an example of a different nature we take a unit impulse occurring at $t = 1$. Thus $V(t) = \delta(t-1)$. Then

$$\begin{aligned} H(f) &= \int_{-\infty}^{\infty} \delta(t-1)\,\mathrm{cas}\,2\pi f t\,dt \\ &= \mathrm{cas}\,2\pi f. \end{aligned}$$

This result follows from use of the sifting property of the impulse function, viz. $\int_{-\infty}^{\infty} \delta(t-a)\phi(t)\,dt = \phi(a)$. In other words, integration of the product of a unit impulse with a given function wipes out all trace of the original function values at all points except where the impulse is and sifts out the function value at the point where the impulse is located.

The sifting theorem may be taken as a point of departure for the theory of impulses as in Schwartz's mathematical development [L. Schwartz, *La Théorie des Distributions* (Herman, 1950)]; or the theory may be deduced as a consequence of a limiting operation inherited from the history of physics. The mathematical rules summarizing the physics approach, which is where impulses originated, are three.

1. Replace $\delta(\cdot)$ by $\tau^{-1}\Pi(\cdot/\tau)$. [In case the dot in $\delta(\cdot)$ stands for a cluster of symbols, the second dot stands for exactly the same cluster. This enables us to discuss $\delta(2t), \delta(t-t_0), \delta(\omega t - \phi)$ and in general $\delta[f(t)]$.]

2. Carry out the integration, or other indicated operation, which should be easy because the factor $\Pi(\cdot/\tau)$ can possess only two values, 0 and 1.

3. Take the limit as $\tau \to 0$ of the result of step 2.

This limit is the value to be assigned to the integral or other expression containing $\delta(\cdot)$. Let us apply these rules to an integral which in the first instance we evaluated by use of the sifting theorem:

$$\int_{-\infty}^{\infty} \delta(t - \tfrac{1}{2}) \operatorname{cas} 2\pi f t \, dt.$$

Applying rule 1 gives

$$\begin{aligned} I &= \int_{-\infty}^{\infty} \tau^{-1} \Pi\left(\frac{t - \frac{1}{2}}{\tau}\right) \operatorname{cas} 2\pi f t \, dt \\ &= \tau^{-1} \int_{\frac{1}{2}-\frac{1}{2}\tau}^{\frac{1}{2}+\frac{1}{2}\tau} \Pi\left(\frac{t - \frac{1}{2}}{\tau}\right) \operatorname{cas} 2\pi f t \, dt. \end{aligned}$$

The limits of integration can be changed as shown because $\Pi[(t - \frac{1}{2})/\tau]$ is centered on $t = \frac{1}{2}$, has a width τ and is thus nonzero only between $\frac{1}{2} - \frac{1}{2}\tau$ and $\frac{1}{2} + \frac{1}{2}\tau$. Now to continue with rule 2,

$$\begin{aligned} I &= \tau^{-1} \int_{\frac{1}{2}-\frac{1}{2}\tau}^{\frac{1}{2}+\frac{1}{2}\tau} \operatorname{cas} 2\pi f t \, dt \\ &= \frac{\Big[\sin 2\pi f t - \cos 2\pi f t\Big]_{\frac{1}{2}-\frac{1}{2}\tau}^{\frac{1}{2}+\frac{1}{2}\tau}}{2\pi f \tau} \\ &= \frac{2 \sin \pi f \tau (\cos \pi f + \sin \pi f)}{2\pi f \tau}. \end{aligned}$$

Finally, using rule 3, we take the limit as $\tau \to 0$ to obtain a value $\cos \pi f + \sin \pi f$ as before. Obviously to know the sifting theorem is faster but the general approach by the three rules always works and is usually easy.

Critically damped pulse

The waveform $V(t) = te^{-t}\mathrm{H}(t)$ is familiar as the impulse response of a critically damped resonator, one whose damping is neither so light as to allow overshoot nor so heavy as to delay recovery. In this case (Fig. 2.4)

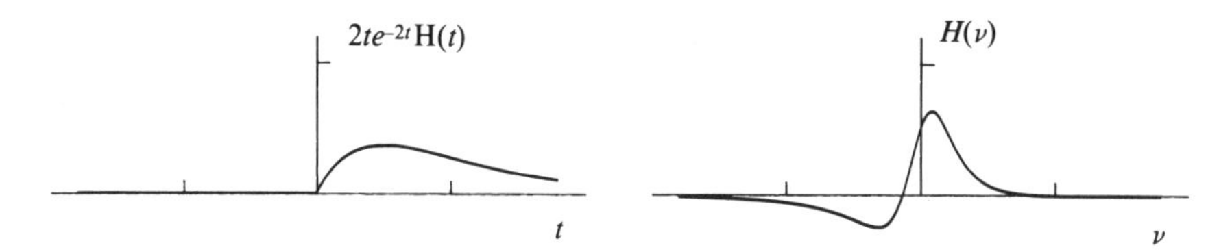

Fig. 2.4. The Hartley transform of the critically damped pulse $2te^{-2t}\mathrm{H}(t)$.

$$\begin{aligned}H(f) &= \int_{-\infty}^{\infty} te^{-t} \operatorname{cas} 2\pi ft\, dt \\ &= \int_{-\infty}^{\infty} t^{-t} \operatorname{cas} 2\pi ft\, dt \\ &= \frac{1 + 4\pi f - 4\pi^2 f^2}{1 + 8\pi^2 f^2 + 16\pi^4 f^4}.\end{aligned}$$

Tables of integrals are helpful in dealing with integration of this kind. Another approach is to deduce the transform by a knowledge of theorems. For example, the transform of $tV(t)$ can be obtained directly if the transform of $V(t)$ is already known. In the next chapter a number of basic theorems are presented for reference.

Power spectrum and phase

It is not always easy to apprehend the run of a complex function from graphs of its real and imaginary parts but the squared modulus or power spectrum is an entity with which one has familiarity from studying optics or other branches of physics. The power spectrum

$$P(f) = |F(f)|^2 = [F_{real}(f)]^2 + [F_{imag}(f)]^2$$

is of necessity an even function and is indeed easier to grasp. On the other hand, of course, one is absorbing only half as much information because the phase has been left out. Nevertheless, the power spectrum may be what one needs in some applications.

To get the power spectrum directly from the Hartley transform is also possible. Thus

$$\begin{aligned}P(f) &= F_{real}^2 + F_{imag}^2 \\ &= E(f)^2 + O(f)^2 \\ &= [H(f) + H(-f)]^2 + [H(f) - H(-f)]^2 \\ &= \frac{[H(f)]^2 + [H(-f)]^2}{2}.\end{aligned}$$

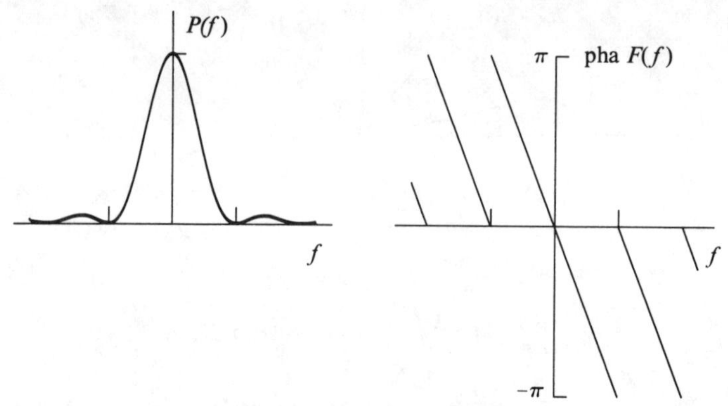

Fig. 2.5. Power spectrum (left) and phase (right) for the gate function $\Pi(t - \frac{1}{2})$.

Thus in lieu of squaring the real and imaginary parts and summing the two values at a given value of f, we square and sum the two values of the Hartley transform at $+f$ and $-f$. Naturally a factor of two is needed because $H(f)$ and $H(-f)$ both have the full quadratic content of the whole spectrum whereas the real and imaginary parts each contain only half the full content.

In optics the phase of the Fourier transform is not easy to measure but in signal analysis phase functions are customary though they require skill to interpret. Examples of power spectrum and phase function are shown in Fig. 2.5. It is quite straightforward to compute the phase from

$$\operatorname{pha} F(f) = \arctan\left[\frac{F_{imag}(f)}{F_{real}(f)}\right].$$

The phase of the Fourier transform is also directly accessible from the Hartley transform. Thus

$$\operatorname{pha} F(f) = \arctan\left[-\frac{H(f) - H(-f)}{H(f) + H(f)}\right] = \arctan\left[-\frac{O(f)}{E(f)}\right].$$

Experience is helpful in interpreting curves of phase versus frequency. The fact is that the interpretation of the phase goes hand in hand with the amplitude; large phase excursions near zero amplitude are not as significant as if the amplitude is large.

To determine the phase of the Fourier transform from the Hartley transform we have

$$\operatorname{pha} F(f) = \arctan\left[\frac{H(-f) - H(f)}{H(f) + H(-f)}\right].$$

A helpful alternative to simultaneous presentation of the real and imaginary parts is to construct a locus on the complex plane by plotting $F_{imag}(f)$ versus $F_{real}(f)$ and to supply the locus with parametric frequency marks. Then at any given frequency the amplitude $|F(f)|$ is the distance of the appropriate point from the origin and the phase is the angular coordinate. Such a diagram is shown in Fig. 2.6 for $\Pi(t-\frac{1}{2})$. The complex plane in this diagram is not the complex plane of complex variable theory, where the independent variable is complex. In this subject the independent variable f is real; but the dependent variable $F(f)$ is complex and one can use its real and imaginary parts as cartesian coordinates.

It is noticeable that, as the locus shrinks in toward the origin, the rate of tracing out of the locus as measured in arc length per frequency interval slows down in such a way that the angular velocity of the tracing point keeps a steady value. This attribute reflects the linear nature of the graph of pha $F(f)$ versus f; the discontinuities in the phase function are seen to be consequences of the passage of the complex locus through the origin.

It is also possible to perceive the transform as a three-dimensional twisted locus of which only a perspective projection can be shown here; but a wire model could be constructed. Fig. 2.7 shows this twisted curve which adds another dimension to one's comprehension. The polar locus can be perceived as the projection of the twisted wire onto the end plane and the orthodox real and imaginary parts are the projections onto the ground plane and side plane respectively.

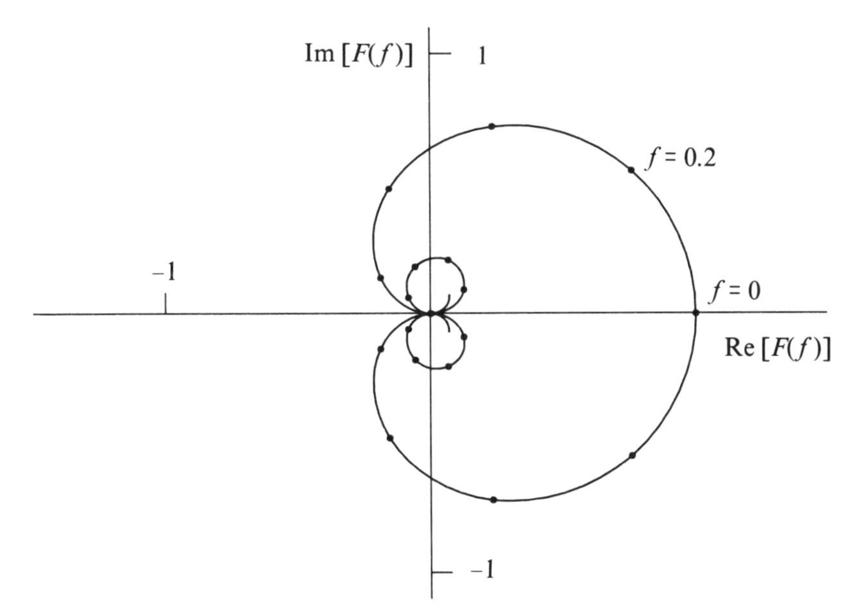

Fig. 2.6. A parametric plot of $F(f)$ on the complex plane where $|F(f)|$ and pha $F(f)$ may be read off as polar coordinates. Parametric marks are at intervals of 0.2 in f.

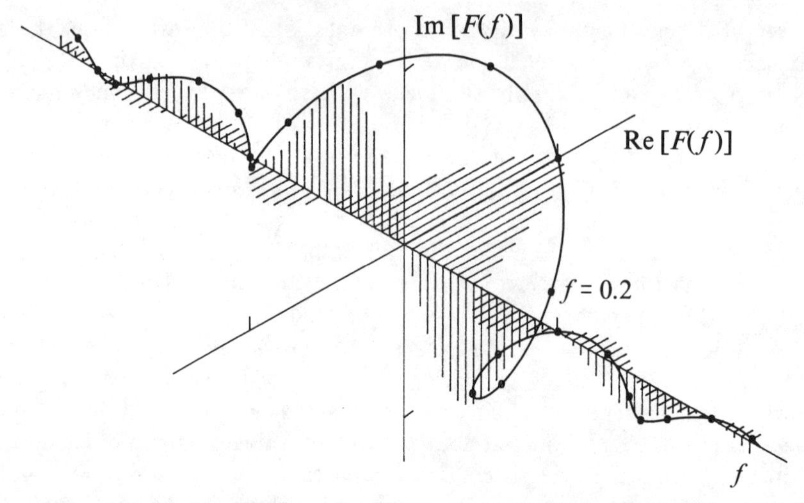

Fig. 2.7. Representation of a complex transform as a twisted curve with parametric marks at intervals of 0.2. The real and imaginary parts are shown ruled.

In a sense the Hartley transform can be seen as an even further form of representation of a real waveform. Being purely real itself, the Hartley transform does not invite diverse modes of presentation. But any of the other modes can be obtained directly from it.

Problems

2.1 *Transform exercises.* Find the Hartley transforms of the following functions. (a) $\sin 2t$, (b) $\cos 2\pi ft$, (c) $e^{-t/2}\mathrm{H}(t)$, (d) $\delta(t+1)$, (e) $(t^2+1)^{-1}$, (f) $\mathrm{H}(t)$, (g) $\mathrm{H}(t-1)$, (h) $\mathrm{H}(t)-\mathrm{H}(t-2)$, (i) $(1-t^2)[\mathrm{H}(t+1)-\mathrm{H}(t-1)]$, (j) $\cos\frac{1}{2}\pi t[\mathrm{H}(t+1)-\mathrm{H}(t-1)]$, (k) t^{-1}, (l) $t\exp(-\pi t^2)\mathrm{H}(t)$.

2.2 *Transform exercises.* Find the Hartley transforms of the following functions. (a) $\cos 3t$, (b) $\sin 2\pi ft$, (c) $e^{-2t}\mathrm{H}(t)$, (d) $\delta(2t+1)$, (e) $(t^2+2)^{-1}$, (f) $t\mathrm{H}(t)$, (g) $2\mathrm{H}(t-2)$, (h) $\mathrm{H}(t-2)-\mathrm{H}(t-3)$, (i) $(1-t^4)[\mathrm{H}(t+1)-\mathrm{H}(t-1)]$, (j) $\cos^2\pi t[\mathrm{H}(t+\frac{1}{2})-\mathrm{H}(t-\frac{1}{2})]$, (k) t^2, (l) $(1-4t^2)[\mathrm{H}(t+\frac{1}{2})-\mathrm{H}(t-\frac{1}{2})]$.

2.3 *Pulse pair.* A waveform $V(t)$ is zero for $t<0$, jumps to a value unity until $t=1$, falls to zero at $t=2$, jumps to unity at $t=3$ and cuts out at $t=4$. What is its Hartley transform?

2.4 *Critically damped pulse.* Find the Hartley transform of $te^{-2t}\mathrm{H}(t)$.

2.5 *Glitch.* A function $V(t)$ is defined to be equal to e^{-t} for positive t and $-e^{t}$ for negative t. Show that its Hartley transform is $4\pi f/(1+4\pi^2 f^2)$.

2.6 *Signum function.* Show that $\operatorname{sgn} t$ has Hartley transform $1/\pi f$.

2.7 *Shah function.* Show that $\mathrm{III}(t)$ has Hartley transform $\mathrm{III}(f)$, where

$$\mathrm{III}(t) \equiv \sum_{n=-\infty}^{\infty} \delta(t-n).$$

2.8 *Hartley transform of complex function.* What is the Hartley transform of the Fourier transform of $V(t)$?

2.9 *Conversion rule.* (a) If $V(t)$ and $H(f)$ are a Hartley transform pair confirm that $V(t)$ and $H(\omega/2\pi)/\sqrt{2\pi}$ satisfy Hartley's original integral. (b) What relationship is satisfied by $V(t/\sqrt{2\pi})$ and $H(\omega/\sqrt{2\pi})$?

2.10 *Complementary function.* Show that $\cos t - \sin t = \operatorname{cas}(-t) = \operatorname{cas}'(t)$.

2.11 *Properties of the* cas *function.* Show that

(a) $\operatorname{cas}(A+B) = \cos A \operatorname{cas} B + \sin A \operatorname{cas}' B$,

(b) $\operatorname{cas}(A-B) = \cos A \operatorname{cas}' B + \sin A \operatorname{cas} B$,

(c) $\operatorname{cas} A \operatorname{cas} B = \cos(A-B) + \sin(A+B)$,

(d) $\operatorname{cas} A + \operatorname{cas} B = 2 \operatorname{cas} \frac{1}{2}(A+B) \cos \frac{1}{2}(A-B)$,

(e) $\operatorname{cas} A - \operatorname{cas} B = 2 \operatorname{cas}' \frac{1}{2}(A+B) \sin \frac{1}{2}(A-B)$.

2.12 *Servomotor.* To drive a motor from one position to another in the minimum time one applies whatever voltage waveform causes the maximum allowable torque. When the motor is halfway to its destination the torque is reversed and when the motor grinds to a halt at its destination the torque is cut. The torque waveform is $\mathrm{H}(t) - 2\mathrm{H}(t-1) + \mathrm{H}(t-2)$. What is the Hartley transform?

2.13 *Single cycle.* As a compromise between speed and power needed, a torque $\sin \pi t[\mathrm{H}(t) - \mathrm{H}(t-2)]$ is substituted for the full ahead, full astern strategy of the previous problem. Determine the Hartley transform and compare with the previous result.

2.14 *Energy conservation.* It is known that if $e(t)$ and $o(t)$ are the even and odd parts of a function $f(t)$ then

$$\int_{-\infty}^{\infty} [f(t)]^2 = \int_{-\infty}^{\infty} [e(t)]^2\, dt + \int_{-\infty}^{\infty} [o(t)]^2\, dt.$$

Is this a unique decomposition? Find, or show the nonexistence of, other dissections of $f(t)$ that obey the same sum rule. Exclude the case where, at any given t, one of the parts is always zero.

CHAPTER 3

THEOREMS

"... to auoide the tediouse repitition of these woordes: is equalle to: I will sette as I doe often in woorke use, a pair of paralleles or Gemowe lines of one lengthe, thus: ══, bicause noe .2. thynges, can be moare equalle."

Robert Recorde, 1557. [First known use of the equality sign.]

Theorems for transforms are helpful for avoiding onerous mathematical analysis in much the same way that the rules of calculus save one the trouble of repeating what has already been done. Equipped with a handful of theorems one may deduce new transforms from old, reduce a given problem to a known one, and combine functions into more elaborate forms without needing to start at the beginning. Integration of functions that are described analytically is a task that is simplified in this way.

Numerical computation also benefits when a knowledge of theorems is brought to bear that converts an operation into a simpler or faster equivalent. Finally, the material provides indispensable thinking tools.

Two sorts of theorem are discussed. The first pertains to operations such as truncation, modulation, convolution and other common operations that may be carried out on a function. This sort of theorem tells what corresponding operation goes on simultaneously in the transform domain. For example, if you simply reverse a waveform by playing it backwards, what happens to its transform? The answer is that the transform is reversed also, which may seem to be so simple as to be trivial. Yet experience shows that prior knowledge of this kind is very helpful, especially where thinking about symmetry can be applied, as in this example.

The second kind of theorem deals with relations between functions and their transforms that can typically be expressed in the form of an equation. For example, the infinite integral of a function is equal to the central value of its transform. Here again is an extremely simple theorem but one that circumvents the need for much integration, is helpful in numerical checking, and is a powerful thing to know about a problem while one is deciding what to analyze or what to compute.

Much of the knowledge about theorems can be condensed into tables that are of permanent value.

Corresponding operations

If $V(t)$ has a Hartley transform $H(f)$, then what will be the transform of $V(t/T)$, i.e. the waveform that results when the time scale is expanded by a factor T? By direct evaluation of the integral, when T is positive,

$$\int_{-\infty}^{\infty} V(t/T)\operatorname{cas} 2\pi f t\, dt = \int_{-\infty}^{\infty} V(t')\operatorname{cas}(2\pi f T t')T\, dt'$$
$$= T\int_{-\infty}^{\infty} V(t')\operatorname{cas}[2\pi(Tf)t']\, dt'$$
$$= TH(Tf).$$

If T is negative then the limits of integration for the new variable $t' = t/T$ are interchanged, a fact that turns the result into $-TH(Tf)$. To accommodate both possibilities we may enunciate the conclusion as follows.

If $V(t)$ has HT $H(f)$ then $V(t/T)$ has HT $|T|H(Tf)$.

By comparison, the Similarity Theorem for the Fourier transform reads:

If $V(t)$ has FT $F(f)$ then $V(t/T)$ has FT $|T|F(Tf)$.

Because of this very close analogy it is convenient to list the theorems for both transforms so that the differences, such as they are, may easily be referred to.

Table 3.1 summarizes the relations. Where the derivations require only one or two lines, as in the foregoing example, they will be omitted.

Convolution

In Table 3.1 convolution is abbreviated by use of the asterisk $*$ and cross-correlation by the pentagram $\star$. With this notation

$$V_1(t) * V_2(t) = \int_{-\infty}^{\infty} V_1(t-u)V_2(u)\, du$$

and

$$V_1(t) \star V_2(t) = \int_{-\infty}^{\infty} V_1(t+u)V_2(u)\, du.$$

An important feature of the convolution theorem is this: if one or both of the functions entering into the convolution is either even or odd, then the Hartley theorem is the same as the Fourier theorem.

If $V_1(t)$ is even then $V_1(t) * V_2(t)$ has HT $H_1(f)H_2(f)$.

And if one of the functions is odd there is a corresponding simplification.

If $V_1(t)$ is odd then $V_1(t) * V_2(t)$ has HT $H_1(f)H_2(-f)$.

Table 3.1 Theorems for the Fourier and Hartley transforms

Theorem	$V(t)$	$F(f)$	$H(f)$
Similarity	$V(t/T)$	$\lvert T\rvert F(Tf)$	$\lvert T\rvert H(Tf)$
Addition	$V_1(t)+V_2(t)$	$F_1(f)+F_2(f)$	$H_1(f)+H_2(f)$
Reversal	$V(-t)$	$F(-f)$	$H(-f)$
Shift	$V(t-T)$	$e^{-i2\pi Tf}F(f)$	$\sin 2\pi Tf\ H(-f)+\cos 2\pi Tf\ H(f)$
Modulation	$V(t)\cos 2\pi f_0 t$	$\frac{1}{2}F(f-f_0)+\frac{1}{2}F(f+f_0)$	$\frac{1}{2}H(f-f_0)+\frac{1}{2}H(f+f_0)$
Convolution	$V_1(t)*V_2(t)$	$F_1(f)F_2(f)$	$\frac{1}{2}[H_1(f)H_2(f)-H_1(-f)H_2(-f)$ $+H_1(f)H_2(-f)+H_1(-f)H_2(f)]$
Autocorrelation	$V(t)\star V(t)$	$\lvert F(f)\rvert^2$	$\frac{1}{2}[H(f)^2+H(-f)^2]$
Product	$V_1(t)V_2(t)$	$F_1(f)*F_2(f)$	$\frac{1}{2}[H_1(f)*H_2(f)+H_1(-f)*H_2(f)$ $+H_1(f)*H_2(-f)-H_1(-f)*H_2(-f)]$
Derivative	$V'(t)$	$i2\pi fF(f)$	$-2\pi fH(-f)$
2nd derivative	$V''(t)$	$-4\pi^2f^2F(f)$	$-4\pi^2f^2H(f)$

Table 3.2 Theorems for relations between domains

Theorem	Property	Fourier relation	Hartley relation
Infinite integral	$\int_{-\infty}^{\infty} V(t)\,dt$	$= F(0)$	$= H(0)$
Rayleigh's	$\int_{-\infty}^{\infty} [V(t)]^2$	$= \int_{-\infty}^{\infty} F(f)F^*(f)\,df$	$= \int_{-\infty}^{\infty} H(f)H^*(f)\,df$
First moment	$\int_{-\infty}^{\infty} tV(t)\,dt$	$= F'(0)/(-i2\pi)$	$= -H'(0)/2\pi$
Second moment	$\int_{-\infty}^{\infty} t^2V(t)\,dt$	$= -F''(0)/4\pi^2$	$= -H''(0)/4\pi^2$
Centroid	$\dfrac{\int_{-\infty}^{\infty} tV(t)\,dt}{\int_{-\infty}^{\infty} V(t)\,dt}$	$= iF'(0)/2\pi F(0)$	$= -H'(0)/2\pi H(0)$

Relations between domains

In addition to theorems of the kind tabulated above, which show what happens to the transform when various operations are applied to the original functions, there are also a number of relations between parameters in the function and transform domains respectively. These relations answer a type of question that frequently arises. Suppose that the transform of a function is available but not the function itself. We desire to know some property of the function but not necessarily the whole function in its entirety. One way would be to invert the transform to get the whole function and then evaluate the desired property. But since requirements are more modest there ought to be a better way. Thus if we wanted the infinite integral of the function we could invert the transform and then integrate the function but the result would be the same as one could get by simply reading off the central value of the transform. Likewise, if we wanted the centroid of a waveform, perhaps in connection with a minimum-phase problem, it would be straightforward to compute by evaluating the ratio of two integrals in accordance with the definition; but how much better it is to know that the abscissa of the centroid is instantly available from the central slope of the transform. Such knowledge, which is obviously very powerful, is condensed in Table 3.2.

Problems

3.1 *Shift theorem.* Derive the shift theorem from the corresponding theorem for the Fourier transform.

3.2 *Autocorrelation theorem.* Show that $V(t) \star V(t)$ has HT $[H_e(f)]^2 + [H_o(f)]^2$, where H_e and H_o are the even and odd parts of $H(f)$.

3.3 *Product theorem.* Show that $V_1(t)V_2(t)$ has HT $H_{1e} * H_{2e} - H_{1o} * H_{2o} + H_{1e} * H_{2o} + H_{1o} * H_{2e}$.

3.4 *Derivative theorem.* Deduce the derivative theorem for the Hartley transform from the knowledge that $V'(t)$ has FT $i2\pi fF(f)$.

3.5 *Derivative theorem.* Obtain the derivative theorem from the shift theorem in the limit as the shift approaches zero.

3.6 *Derivative theorem.* Show that the nth derivative $V^{(n)}(t)$ has HT $\operatorname{cas}'(n\pi/2)(2\pi f)^n H[(-1)^n f]$.

3.7 *Second derivative.* A student reasons as follows. "When you differentiate $V(t)$ the effect on $H(f)$ is to reverse $H(f)$ back to front and multiply by $-2\pi f$. Therefore, if you differentiate again you reverse $H(f)$ again, which takes it back to where it was at the start, and multiply by $-2\pi f$ again. Consequently $V''(t)$ has HT $(-2\pi f)^2 H(f)$." Where is the error in this reasoning?

3.8 *Second moment.* Show that the second moment of a waveform can be determined from its transform as follows: $\int_{-\infty}^{\infty} t^2 V(t)\,dt = -H''(0)/4\pi^2$.

3.9 *Running means.* A smoothed waveform $V_{sm}(t)$ is constructed from $V(t)$ by integrating from $t - \frac{1}{2}$ to $t + \frac{1}{2}$. Show that the Hartley transform of $V_{sm}(t)$ is $H(f)\operatorname{sinc} f$.

3.10 *FT from HT.* Deduce that

$$F(f) = \tfrac{1}{2}e^{-i\pi/4}H(f) + \tfrac{1}{2}e^{i\pi/4}H(-f).$$

3.11 *HT from FT.* Deduce that

$$H(f) = \tfrac{1}{2}e^{i\pi/4}F(f) + \tfrac{1}{2}e^{-i\pi/4}F^*(-f).$$

3.12 *Ramp step.* A function $f(t)$ is defined by

$$f(t) = \begin{cases} -1, & t < -1; \\ t, & -1 < t < 1; \\ 1, & t > 1. \end{cases}$$

A student notices that $g(t) = \Pi(t) * \operatorname{sgn}(t)$ is similar in character but is compressed by a factor 2. Knowing that $g(t) = \Pi(t) * \operatorname{sgn}(t)$ has HT $G(f) = \operatorname{sinc}(f)/2\pi f$, the student applies the similarity theorem to deduce that $f(t) = g(t/2)$ has HT $2G(2f) = \operatorname{sinc}(2f)/\pi f$. Another student reasons as follows: "The given $f(t) = \frac{1}{2}\Pi(t/2) * \operatorname{sgn}(t/2)$; now $\Pi(t) * \operatorname{sgn}(t)$ has HT $G(f)$, therefore $\Pi(t/2) * \operatorname{sgn}(t/2)$ has HT $2G(2f)$. But $\operatorname{sgn}(t/2)$ is the same as $\operatorname{sgn}(t)$; therefore $\frac{1}{2}\Pi(t/2) * \operatorname{sgn}(t) = f(t)$ has HT $G(2f)$." Explain the disagreement.

3.13 *Shifting an even function.* If $f(x)$ is an even function, show that $f(x-a)$ has HT $\operatorname{cas}(2\pi f)H(f)$.

CHAPTER 4

THE DISCRETE HARTLEY TRANSFORM

"The intellect of man is forced to choose
Perfection of the life or of the work."

W.B. Yeats

Although we tend to think of time as a continuous variable it is necessary in practice to use a discrete variable to describe time series, for example when a computer requires discretization of the variable or when data are accumulated at regular intervals. We therefore introduce a discrete variable τ which may be thought of as relating to time but which can assume only integral values ranging from 0 to $N-1$. This range is chosen rather than 1 to N or $-\frac{1}{2}N+1$ to $\frac{1}{2}N$ to accord with other practice. Thus the Discrete Fourier Transform (DFT) and its inversion integral have the standard form

$$F(\nu) = N^{-1} \sum_{\tau=0}^{N-1} f(\tau) \exp(-i2\pi\nu\tau/N)$$

$$f(\tau) = \sum_{\nu=0}^{N-1} F(\nu) \exp(2\pi\nu\tau/N).$$

The function $f(\tau)$ may be the discrete representation of an underlying continuous waveform or may be a function of a variable that is basically discrete.

The discrete Hartley transform integral

The Discrete Hartley Transform (DHT) of a real function $f(\tau)$ is, with its inverse, given by

$$H(\nu) = N^{-1} \sum_{\tau=0}^{N-1} f(\tau)\text{cas}(2\pi\nu\tau/N)$$

$$f(\tau) = \sum_{\nu=0}^{N-1} H(\nu)\text{cas}(2\pi\nu\tau/N)$$

where, as before, we use the abbreviation $\text{cas}\,\theta = \cos\theta + \sin\theta$ taken from Hartley.

To derive the inverse DHT integral we use the orthogonality relation

$$\sum_{\nu=0}^{N-1} \text{cas}(2\pi\nu\tau/N)\,\text{cas}(2\pi\nu\tau'/N) = \begin{cases} N, & \tau = \tau' \\ 0, & \tau \neq \tau'. \end{cases}$$

Substituting $N^{-1}\sum_{\tau=0}^{N-1} f(\tau)\text{cas}(2\pi\nu\tau/N)$ for $H(\nu)$ in the expression $\sum_{\nu=0}^{N-1} H(\nu)\text{cas}(2\pi\nu\tau/N)$, we have

$$\begin{aligned}
\sum_{\nu=0}^{N-1} H(\nu)\text{cas}(2\pi\nu\tau/N) &= \sum_{\nu=0}^{N-1} N^{-1} \sum_{\tau'=0}^{N-1} f(\tau')\text{cas}(2\pi\nu\tau'/N)\text{cas}(2\pi\nu\tau/N) \\
&= N^{-1} \sum_{\tau'=0}^{N-1} f(\tau') \sum_{\nu=0}^{N-1} \text{cas}(2\pi\nu\tau'/N)\text{cas}(2\pi\nu\tau/N) \\
&= N^{-1} \sum_{\tau'=0}^{N-1} f(\tau') \times \begin{cases} N, & \tau = \tau' \\ 0, & \tau \neq \tau' \end{cases} \\
&= f(\tau),
\end{aligned}$$

which verifies the inversion integral.

The factor N^{-1} in the DHT is borrowed from DFT practice where $F(0)$ is the d.c. value of $f(\tau)$; otherwise the DHT transformation is symmetrical. In addition the DHT is real, because $f(\tau)$ is real.

Meaning of τ and ν

Just as τ reminds us of time, so the discrete variable ν reminds us of frequency; however, there are two points to remember. If the unit of t is the second, that is if the time interval between successive elements of the time series $f(\tau)$ is one second, then it is ν/N, not ν, that will be measured in hertz. The frequency interval between successive elements of the sequence $H(\nu)$ is N^{-1} Hz. As ν increases the corresponding frequency increases, but only as far as $\nu = N/2$; beyond that the frequency corresponding to ν is $(N-\nu)/N$, reaching zero where $\nu = N$.

Even and odd parts

As in the continuous case, the DHT possesses even and odd parts

$$H(\nu) = E(\nu) + O(\nu)$$

but some thought has to be given to the definitions because of the convention that ν should range from 0 to $N-1$. The conventional way of handling this is to assign function values outside the basic range so as to generate a cyclic function with period N. So at $\nu = -1$ we assign $H(N-1)$ because $\nu = -1$ and $\nu = N-1$ are separated by one period of length N. In general we assign to $H(-\nu)$, where $-N \leq \nu \leq -1$, the value $H(N-\nu)$, for which the independent variable is in the basic range of ν. By this procedure we achieve a simpler relationship between ν and frequency: we can say that ν/N is the same as frequency in Hz, over the range $-N/2 < \nu < N/2$. We also arrive at equations for the even and odd parts that are consistent with what has gone before. Thus

$$E(\nu) = \frac{[H(\nu) + H(N-\nu)]}{2}$$

and

$$O(\nu) = \frac{[H(\nu) - H(N-\nu)]}{2}.$$

From the definition of the DFT $F(\nu)$ it is apparent that $F(\nu)$ can be formed from the DHT's even and odd parts by

$$F(\nu) = E(\nu) - iO(\nu).$$

Conversely, to form $H(\nu)$ when $F(\nu)$ is available

$$H(\nu) = \operatorname{Re} F(\nu) - \operatorname{Im} F(\nu).$$

These relations are strictly analogous to those obtained previously for the continuous variable.

Examples of DHTs

A number of examples will be given to illustrate the character of the DHT. First some analytic expressions will be taken and then some sequences specified numerically.

Exponential tail

For comparison with the continuous example given in an earlier chapter consider

$$f(\tau) = \begin{cases} 0.5, & \tau = 0; \\ \exp(-\tau/2), & \tau = 1, 2, \ldots 15, \end{cases}$$

which represents the function of continuous t that was used before by $N = 16$ equispaced samples. The value at $\tau = 0$, since it falls on the discontinuity of $V(t)$, is assigned the value $[V(0+)+V(0-)]/2 = 0.5$. The result for $H(\nu)$, which is shown in Fig. 4.1, closely resembles samples of the continuous function taken at intervals $\Delta\omega/2\pi = 1/16$.

The discrepancies, which are small in this example, are due partly to the truncation of the exponential waveform and partly to aliasing, exactly as with the DFT.

Binomial pulse

For a smoother example, take the binomial sequence 1, 6, 15, 20, 15, 6, 1 representing samples of a smooth pulse. To obtain the simplest result

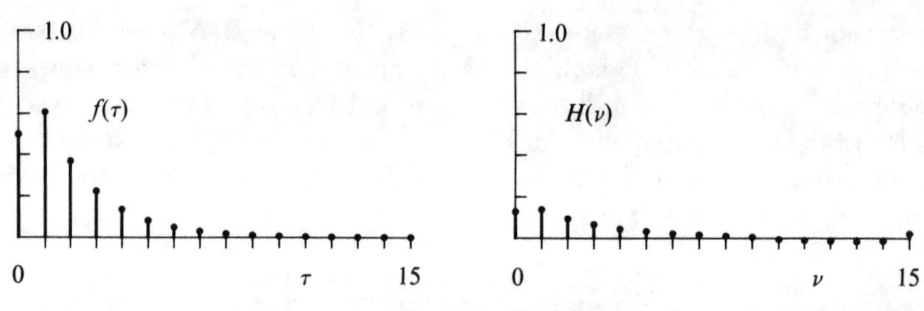

Fig. 4.1. A 16-point representation of the truncated exponential waveform (left) previously used for illustrating the continuous transform and its DHT (right).

assign the peak value at $\tau = 0$. Thus $f(\tau) = 6!/(3-\tau)!(\tau+3)!$ and the result, as shown in Table 4.1 and in Fig. 4.2, is the expected smooth pulse peaking at $\nu = 0$.

For numerical checking it is useful to know that, as for the discrete Fourier transform, the sum of the DHT values $\Sigma H(\nu)$ is equal to $f(0)$. Conversely, the sum of the data values $\Sigma f(\tau)$ is equal to $NH(0)$. Thus if

$$\{f(0) \quad f(1) \quad f(2) \quad f(3)\} \quad \text{has DHT} \quad \{H(0) \quad H(1) \quad H(2) \quad H(3)\},$$

then
$$f(0) = H(0) + H(1) + H(2) + H(3)$$
and
$$f(0) + f(1) + f(2) + f(3) = 4 \times H(0).$$

Gate function

Consider a gating operator that selects the second group of four elements from a set of 16 and replaces the remaining 12 by zeros; this operation is equivalent to multiplying by

$$\{0\ 0\ 0\ 0 \quad 1\ 1\ 1\ 1 \quad 0\ 0\ 0\ 0 \quad 0\ 0\ 0\ 0\}.$$

The DHT $H_1(\nu)$ is shown in Fig. 4.3. If now the different gating operator

$$\{1\ 1\ 0\ 0 \quad 0\ 0\ 0\ 0 \quad 0\ 0\ 0\ 0 \quad 0\ 0\ 1\ 1\}$$

is considered there is quite a difference in the DHT $H_2(\nu)$ as shown in Fig. 4.4. Numerical results for both cases, normalized to a leading value of unity, are listed in Table 4.2.

In the second case the gating is symmetrical about $\tau = 0$, as is brought

Table 4.1 A binomial series and its DFT

$\tau, \nu =$	0	1	2	3	4	5	6	7	8	9	10	11	12	13	14	15
$f(\tau) =$	20	15	6	1	0	0	0	0	0	0	0	0	0	1	6	15
$H(\nu) =$	4	3.56	2.49	1.32	0.5	0.12	0.01	0	0	0	0.01	0.12	0.5	1.32	2.49	3.56

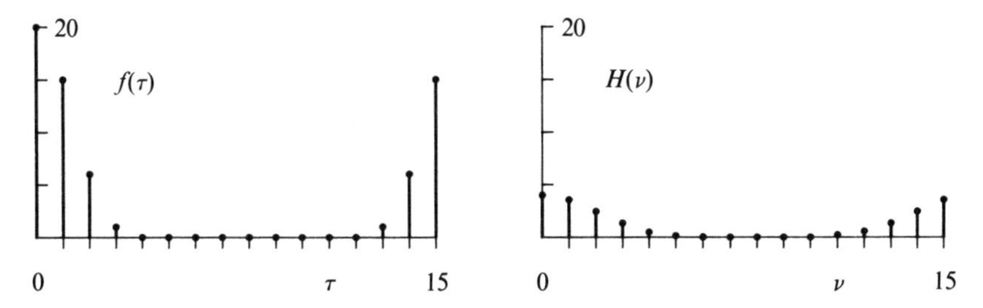

Fig. 4.2. Samples representing a smooth binomial hump (left) and its DHT (right).

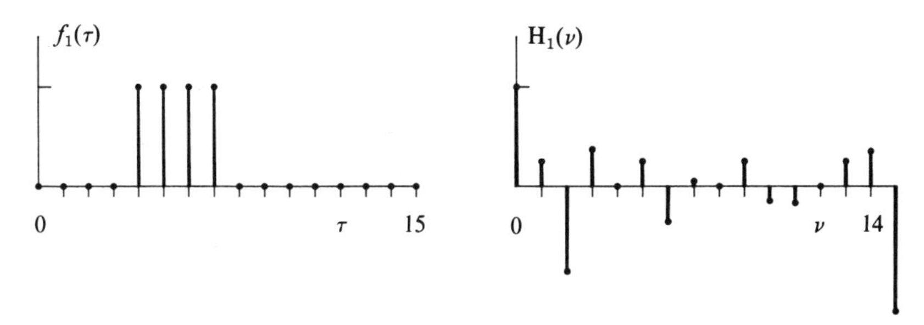

Fig. 4.3 The DHT $H_1(\nu)$ of a gate function.

Table 4.2 The DHTs of two gate functions

τ, ν	$f_1(\tau)$	DHT_1	$f_2(\tau)$	DHT_2
0	0	1	1	1
1	0	0.25	1	0.853
2	0	−0.854	0	0.483
3	0	0.374	0	0.07
4	1	0	0	−0.2
5	1	0.25	0	−0.236
6	1	−0.354	0	−0.083
7	1	0.05	0	0.1135
8	0	0	0	0.2
9	0	0.25	0	0.1135
10	0	−0.146	0	−0.083
11	0	−0.167	0	−0.236
12	0	0	0	−0.2
13	0	0.25	0	0.07
14	0	0.354	1	0.483
15	0	−1.257	1	0.8525

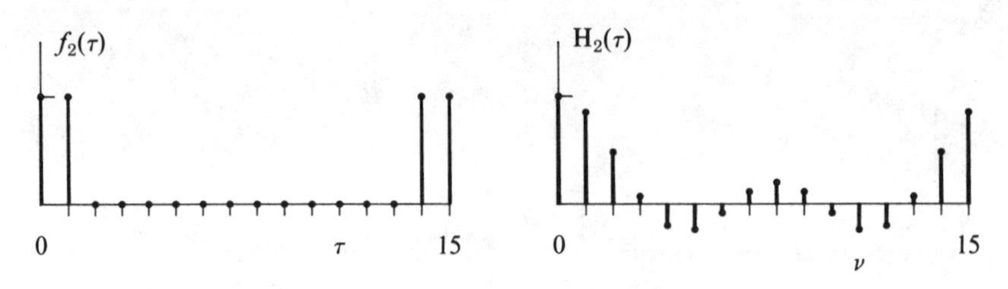

Fig. 4.4 The DHT $H_2(\nu)$ of a gate function centered on $\tau = 0$.

out by thinking of the cyclic representation. Consequently there is freedom from fast oscillation and the result *resembles* the sinc function associated with a central rectangle function.

Degrees of freedom

One to one relationships have been established between the discrete Fourier transform and the discrete Hartley transform. This raises a question of information theory. How does one explain that N real values of the DHT can substitute for the N complex values of the DFT, a total of $2N$ real numbers? This may be understood by remembering that the Hermitian property of the DFT means redundancy by a factor of two; thus the DFT has only N degrees of freedom even though there are $2N$ real coefficients. Since there is no degeneracy due to symmetry in the DHT the N real coefficients are equivalent to the N complex coefficients of the DFT.

Other real kernels

The function $\operatorname{cas}\theta$, which may be thought of as a sine wave shifted 45^0, automatically responds to cosine and sine components equally. If, as a kernel, we use $2^{1/2}\sin(\theta+\alpha)$, where α is an arbitrary shift, the responses will be unequal but no information will be lost unless $\alpha = 0, \pi/2, \ldots$. Consequently one would expect to be able to invert; the inversion kernel is $\cot^{1/2}\alpha\sin\theta + \tan^{1/2}\alpha\cos\theta$.

Theorems

Just as there is a Hartley transform theorem for every theorem that applies to the Fourier transform, so also there are corresponding theorems for the discrete transforms. In the case of convolution and autocorrelation, theorems are included in Tables 4.3 and 4.4 for completeness but they are not discussed until the next chapter. It will be sufficient to make a note here that convolution of functions of a continuous variable as represented by the symbol $*$ differs from cyclic convolution of discrete sequences as represented by $\circledast$.

One may note that the mean value of the sequence $f(\tau)$ is given by $H(0)$ and that the mean square value of $f(\tau)$ is given by ΣH^2.

Some of the theorems for the two different transforms correspond exactly, as is the case with $\sum H(\nu) = f(0)$ and $\sum f(\tau) = N \times H(0)$, but some exhibit differences.

Table 4.3 Theorems for operations on discrete transforms

Theorem	Function $f(\tau)$	DFT $F(\nu)$	DHT $H(\nu)$
Reversal	$f(-\tau)$	$F(-\nu)$	$H(-\nu)$
Addition	$f_1(\tau)+f_2(\tau)$	$F_1(\nu)+F_2(\nu)$	$H_1(\nu)+H_2(\nu)$
Shift	$f(\tau-T)$	$e^{-i2\pi T\nu/N}F(\nu)$	$\cos(2\pi T\nu/N)H(\nu)$ $-\sin(2\pi T\nu/N)H(N-\nu)$
Convolution	$f_1(\tau)\circledast f_2(\tau)$	$NF_1(\nu)F_2(\nu)$	$\frac{1}{2}N[H_1H_2-H_1(-)H_2(-)$ $+H_1H_2(-)+H_1(-)H_2]$
Product	$f_1(\tau)f_2(\tau)$	$F_1(\nu)\circledast F_2(\nu)$	$\frac{1}{2}N[H_1\circledast H_2-H_1(-)\circledast H_2(-)$ $+H_1\circledast H(-)+H_1(-)\circledast H_2]$
Autocorrelation	$f(\tau)\circledast f(\tau)$	$\frac{1}{2}N\,\lvert F(\nu)\rvert^2$	$N\{[H(\nu)]^2+[H(-\nu)]^2\}$
Stretch	$\{f(0)\ 0\ f(1)...\}$		*See text*
Derivative	$f'(\tau)$	$i2\pi\nu F(\nu)$	$2\pi\nu H(-\nu)$
2nd derivative	$f''(\tau)$	$-4\pi^2\nu^2F(\nu)$	$-4\pi^2\nu^2H(\nu)$

Reversal theorem

When a sequence $f(\tau)$ is reversed to become $f(-\tau)$ the outcome is that the leading element remains in place because its index is zero and the rest are reversed in order. Thus $f(1)$ becomes $f(-1)$ which, interpreted as $f(-1 \bmod N)$, is $f(N-1)$, or the last element. Consequently $\{a\ b\ c\ d\quad e\ f\ g\ h\}$ reverses into $\{a\ h\ g\ f\quad e\ d\ c\ b\}$ and the transform $\{A\ B\ C\ D\ \ E\ F\ G\ H\}$ reverses into $\{A\ H\ G\ F\ \ E\ D\ C\ B\}$.

Addition theorem

The superposition property evidenced by the addition theorem is simply an expression of the linearity of the DHT operator.

Shift theorem

First take an example where unit shift is applied to a sequence

Table 4.4 Theorems for relations for discrete transforms

Sum of sequence	$\sum_{\tau=0}^{N-1} f(\tau) = NF(0) = NH(0)$
First value	$f(0) = \sum_{\nu=0}^{N-1} F(\nu) = \sum_{\nu=0}^{N-1} H(\nu)$
Quadratic content	$\sum_{\tau=0}^{N-1} [f(\tau)]^2 = N\sum_{\nu=0}^{N-1} \lvert F(\nu)\rvert^2 = N\sum_{\nu=0}^{N-1} \lvert H(\nu)\rvert^2$

$$\{a_0 \quad a_1 \quad a_2 \quad \dots \quad a_{N-1}\}$$

such that

$$\{a_0 \quad a_1 \quad a_2 \quad \dots \quad a_{N-1}\} \quad \text{has DHT} \quad \{\alpha_0 \quad \alpha_1 \quad \alpha_2 \quad \dots \quad \alpha_{N-1}\}.$$

According to the shift theorem, with $T = 1$,

$$\{a_{N-1} \quad a_0 \quad a_1 \quad a_2 \dots a_{N-2}\} \quad \text{has DHT}$$

$$\{\alpha_0 \ C_1\alpha_1 \ C_2\alpha_2 \dots C_{N-1}\alpha_{N-1}\} - \{0 \ S_1\alpha_{N-1} \ S_2\alpha_{N-2} \dots S_{N-1}\alpha_1\},$$

where $C_\nu = \cos(2\pi\nu/N)$ and $S_\nu = \sin(2\pi\nu/N)$.

In order to perform the shift one moves each element one place to the right. The last element is brought back to the beginning in accordance with the cyclic convention.

The DHT is composed of two sequences, one containing cosine coefficients, the other sine coefficients. The DFT of a shifted input is also composed of two sequences with sine and cosine coefficients but the sine sequence for the DHT runs backwards, a feature referred to as retrograde indexing. To derive the shift theorem, substitute $f(t+T)$ in the definition to obtain

$$\begin{aligned}
\sum_{\tau=0}^{N-1} f(\tau+T)\operatorname{cas}(2\pi\nu\tau/N) &= \sum_{\tau'=T}^{N-1+T} f(\tau')\operatorname{cas}[2\pi\nu(\tau'-T)/N] \\
&= \sum_{\tau'} f(\tau')[\operatorname{cas}(2\pi\nu\tau'/N)\cos(2\pi\nu T/N) \\
&\qquad + \operatorname{cas}'(2\pi\nu\tau'/N)\sin(2\pi\nu T/N)] \\
&= \cos(2\pi\nu T/N)\sum_{\tau'} f(\tau')\operatorname{cas}(2\pi\nu\tau'/N) \\
&\qquad + \sin(2\pi\nu T/N)\sum_{\tau'} f(\tau'\operatorname{cas}'(2\pi\nu\tau'/N) \\
&= \cos(2\pi\nu T/N)H(\nu) \\
&\qquad - \sin(2\pi\nu T/N)\sum_{\tau'} f(\tau')\operatorname{cas}(-2\pi\nu\tau'/N) \\
&= \cos(2\pi\nu T/N)H(\nu) - \sin(2\pi\nu T/N)H(-\nu).
\end{aligned}$$

Convolution theorem

In general, the transform of the convolution $f_1(\tau) \circledast f_2(\tau)$ contains four terms. The fundamental quantities to be computed are the direct products $P_a(\nu) = H_1(\nu)H_2(\nu)$ and the retrograde products $P_b(\nu) = H_1(\nu)H_2(-\nu)$. In these terms,

$$H(\nu) = \tfrac{1}{2}N[P_a(\nu) - P_a(-\nu) + P_b(\nu) + P_b(-\nu)].$$

Thus only two multiplications, rather than four, are involved. Now if $H_2(\nu)$ is even (i.e. $H_2(\nu) = H_2(-\nu)$), then

$$H(\nu) = NH_1(\nu)H_2(\nu).$$

A similarly simple form results if $H_2(\nu)$ is odd; in this case

$$H(\nu) = NH_1(-\nu)H_2(\nu).$$

Because of commutativity, $H(\nu) = H_1(\nu)H_2(\nu)$ if *either* $H_1(\nu)$ or $H_2(\nu)$ is even. It often happens that one or other of the functions has symmetry or antisymmetry and then the simpler relations apply. Because of its importance convolution is returned to in a later chapter.

Product theorem

The four terms in the product theorem involve only two multiplications because two of the terms are obtainable merely by reversing the other two.

Stretch theorem

The similarity theorem in the case of the continuous independent variable refers to a change of scale of the abscissa, as when $V(t)$ is changed to $V(t/T)$. As T may be greater or less than unity the operation may be one of stretching or compressing. In the case of a discrete variable changes of scale are also of practical importance as for example when a sequence of regular measurements is to be repeated at a faster or slower rate. But whereas $V(t/T)$ is determinate for any T when $V(t)$ is given, the same is not true of $f(\tau/T)$ when $f(\tau)$ is given for $\tau = 0, 1, \ldots N-1$. Consequently there is no strict analogue for the similarity theorem. The stretch theorem refers to expansion only, and a kind of expansion that is produced by the insertion of zeros. It is most simply presented by an example.

Let

$$\{a \quad b \quad c \quad d\} \quad \text{have DHT} \quad \{\alpha \quad \beta \quad \gamma \quad \delta\}.$$

Then

$$\{a\ 0\ b\ 0 \quad c\ 0\ d\ 0\} \quad \text{has DHT} \quad \tfrac{1}{2}\{\alpha\ \beta\ \gamma\ \delta \quad \alpha\ \beta\ \gamma\ \delta\}.$$

This result may be verified by examining the expression for the DHT of the right hand side: $\frac{1}{2}\alpha + \frac{1}{2}\beta \operatorname{cas} \tau\Theta + \frac{1}{2}\gamma \operatorname{cas} 2\tau\Theta + \frac{1}{2}\delta \operatorname{cas} 3\tau\Theta + \frac{1}{2}\alpha \operatorname{cas} 4\tau\Theta + \frac{1}{2}\beta \operatorname{cas} 5\tau\Theta + \frac{1}{2}\gamma \operatorname{cas} 6\tau\Theta + \frac{1}{2}\delta \operatorname{cas} 7\tau\Theta$. When $\tau = 0$ we verify that $f(0) = a$. When τ is odd the sum is zero. When τ is even the sum reduces to $\alpha + \beta \operatorname{cas} \tau\Theta + \gamma \operatorname{cas} 2\tau\Theta + \delta \operatorname{cas} 3\tau\Theta$, whose DHT is $\{a \quad b \quad c \quad d\}$.

Conclusion

The properties of the DHT commend themselves for application to numerical analysis. The fact that the transform values are real is a convenience in managing calculations. In addition, the reversibility of the transform is helpful as one does not need to keep track of which domain one is in. Furthermore, several of the theorems for the Fourier transform have different forms according to the domain, a concern that is obviated with the DHT. The factor N is domain-dependent and could also be avoided but in practice there is nearly always a normalizing factor or calibration factor to be applied at the end of a numerical calculation. Experience shows that the last step is the place to consolidate proportionality factors and so the departure from strict reversibility represented by the factor N is not computationally important.

Problems

4.1 *Addition theorem.* Given that {1 2 3 4 5 6 7 8} has DHT {4.5 − 1.71 − 1 − 0.71 − 0.5 − 0.29 0 0.71} show how to use the addition theorem to deduce directly the DHT of {8 7 6 5 4 3 2 1}.

4.2 *Reversal theorem.* If {1 2 4 8 16 32 64 128} has DHT

$$\tfrac{1}{8}\{255 \quad -117 \quad -153 \quad -125 \quad -85 \quad -33 \quad 51 \quad 215\},$$

what is the DHT of {1 128 64 32 16 8 4 2}?

4.3 *Shift theorem.* A certain sampled waveform $f(\tau)$ has DHT

$$H(\nu) = \{32 \quad 0 \quad -6 \quad 10 \quad 32 \quad 0 \quad -6 \quad 10\}.$$

What are the DHTs of (a) $f(\tau - 4)$, (b) $f(\tau + 4)$?

4.4 *Shift theorem.* Let $H(\nu) = \{0\ 0.884\ 1.25\ -0.884\ -3.75\ -4.419\ -2.5\ -0.884\}$. What does $H(\nu)$ become if the samples of $f(\tau)$ are taken earlier by one unit of τ?

4.5 *Autocorrelation theorem.* (a) What is the autocorrelation of

$$f(\tau) = \{1\ 1\ 1\ 0 \quad 0\ 0\ 0\ 0 \quad 0\ 0\ 0\ 0 \quad 0\ 0\ 1\ 1\}?$$

(b) What is the DHT of $f(\tau)$? (c) Give the DHT of the autocorrelation of $f(\tau)$.

4.6 *Convolution theorem.* Show how to derive the Hartley transform of a convolution from the known result $F(\nu) = F_1(\nu)F_2(\nu)$ for the Fourier transform.

4.7 *Product theorem.* Prove the product theorem.

4.8 *Autocorrelation theorem.* Show how to deduce the autocorrelation theorem from the convolution theorem.

4.9 *Numerical exercises.* Find the DHT of the following sequences.

(a) {3 1 4 1 5 9 2 6},
(b) {1 4 2 8 5 7 1 4},
(c) {2 7 1 8 2 8 1 8},
(d) {3 1 8 3 0 9 8 8}.

4.10 *Numerical exercises.* Find the DHT of the following sequences.
(a) {2 8 5 7 1 4 2 8},
(b) {4 2 8 5 7 1 4 2},
(c) {5 7 1 4 2 8 5 7},
(d) {7 1 4 2 8 5 7 1},
(e) {8 5 7 1 4 2 8 5}.

4.11 *Binomial sequences.* Find the DHTs of the following sequences.
(a) {1 7 21 35 35 21 7 1},
(b) {0 1 5 10 10 5 1 0},
(c) {20 15 6 1 0 1 6 15},
(d) {35 35 21 7 1 1 7 21}.

4.12 *Special cases.* What are the DHTs of {1 2} and {0.9 0.1}?

4.13 *Trailing zeros.* What are the DHTs of
(a) {3 1 4 1 5 9 2 6 0 0 0 0 0 0 0 0},
(b) {0 0 0 0 3 1 4 1 5 9 2 6 0 0 0 0},
(c) {0 0 0 0 0 0 0 0 3 1 4 1 5 9 2 6}?

4.14 *Sequences of prime length.* Find the DHTs of
(a) {1 2 3},
(b) {1 2 3 4 5},
(c) {1 2 3 4 5 6 7},
(d) {1 2 3 4 5 6 7 8 9 10 11}.

4.15 *Random sequences.* The elements of a sequence have values of ± 1 at random. (a) Explain why $H(0)$ is approximately zero. (b) What do you expect of $H(N/2)$? (c) What would be the standard deviation σ of $H(0)$ for numerous random sequences?

4.16 *Integer transforms.* An 8-element sequence consists of integers only. Under what condition will the DHT consist only of integers?

4.17 *Compare DHT with HT.* Obtain the DHT of {8 7 6 5 4 3 2 1 0 0 0 0 0 0 0 0 0 1 2 3 4 5 6 7} and compare with $(64/24)\,\mathrm{sinc}^2(8\nu/24)$.

CHAPTER 5

DIGITAL FILTERING BY CONVOLUTION

"I saw in a flash that 987654321 by 81 equals 80000000001, and so I multiplied 123456789 by this, a simple matter, and divided the answer by 81. Answer 12193263111263 5269. The whole thing can hardly have taken more than half a minute."

R.J. Aitken, Biographical Memoirs of Fellows of the Royal Society, 1968.
[Aitken multiplying 987654321 by 123456789 on request from his children.]

Filtering is a general term from electrical engineering to describe the selection of wanted frequency bands from a signal; and while one may think of a filter as a device that operates on the spectrum by transmitting certain frequency bands and blocking others, it is equally valid to think of a filter as operating on the waveform in the time domain to modify it in the desired way. The same duality exists in computing. A noisy stream of data from which a band of noise has to be removed may first be converted to a spectral representation, components at the unwanted frequencies may be replaced by zeros, and then one may return to the data domain by inverting the transformation. But alternatively, the same result may be achieved directly in the data domain merely by convolving the data with a suitable filtering sequence without thinking about frequency at all.

Meteorological data from which seasonal fluctuations have to be removed in order to display secular climatic trends have traditionally been subjected to running means by convolving the sequence of monthly values with the 13-element sequence $\frac{1}{12}\{\frac{1}{2}\ 1\ 1\ 1\ 1\ 1\ \underline{1}\ 1\ 1\ 1\ 1\ 1\ \frac{1}{2}\}$, an operation that may be described as low-pass filtering. If one subtracts the smoothed trend from the original monthly sequence one obtains the fluctuations themselves, free from the slow drift from year to year. This could be described as low-stop filtering and could be carried out in one step, if desired, by convolution with $\frac{1}{12}\{-\frac{1}{2}\ -1\ -1\ -1\ -1\ -1\ +\underline{1}\ -1\ -1\ -1\ -1\ -1\ -\frac{1}{2}\}$. These two examples show how the frequency spectrum of a signal may in effect be operated upon by direct convolution in the data domain.

Just as an electronic filter, such as the combination of electrolytic capacitors, resistors and iron-cored inductors used in a television set to reduce power-frequency hum, does not succeed in entirely removing the unwanted signal, so it is also with computed filtering on data streams. To understand the performance and limitations of numerical filtering is an important goal.

Interestingly, the direct way of filtering by convolution may not be the fastest. Thus the choice of procedure is another technical topic. When the data stream is sufficiently lengthy it *may* pay to take the transform and work in the frequency domain. But there is another consideration, namely the length of the convolving sequence; would it pay to transform a string of 1000 monthly values which have to be smoothed, operate on the spectrum and then retransform, rather than to carry out the rather simple multiplying and adding called for by the conventional 13-element running-means sequence?

This and other questions will be addressed in this chapter.

Cyclic convolution

The ordinary convolution sum of two sequences $f_1(\tau)$ and $f_2(\tau)$ is defined by

$$f_1(\tau) * f_2(\tau) = \sum_{\tau'} f_1(\tau') f_2(\tau - \tau').$$

In this definition τ' ranges over the domain of nonzero products for each given τ, and τ itself may range to infinity. The two sequences may have any length from one to infinity and the two lengths need not be the same. If the lengths are N_1 and N_2 then $f_1 * f_2$ has length $N_1 + N_2 - 1$. The number of values assumed by the dummy index τ' when the products are being scanned depends on the current selection of τ; the number ranges from a low of one up to the lesser of N_1 and N_2. A program for ordinary convolution performed in accordance with the defining summation is listed in Appendix 1 under the name **CONV**.

In cyclic convolution on the other hand both sequences are defined on a circle of perimeter N. Thus the variable τ may be thought of as ranging round and round the circle but as having only N distinctive values. There are exactly N products entering into each of the N output values. The cyclic convolution $f_{cyc}(\tau)$ is defined by

$$f_{cyc}(\tau) = \sum_{\tau'=0}^{N-1} f_1(\tau' \bmod N) f_2[(\tau - \tau') \bmod N].$$

In this expression τ' is shown as running from 0 to $N - 1$ but it would make no difference if τ' ran from 1 to N; it is only necessary that all N products be summed. In thinking about convolution many people regard the leading element in a data sequence as the first rather than the zeroth and this understandable attitude should not cause trouble.

As an example of cyclic convolution consider Fig. 5.1 which represents the case where

$$\begin{aligned} f_1(\tau) &= \{1\,1\,1\,1 \quad 1\,1\,1\,0 \quad 0\,0\,0\,0 \quad 0\,0\,0\,0\} \\ f_2(\tau) &= \{1\,1\,1\,0 \quad 0\,0\,0\,0 \quad 0\,0\,0\,0 \quad 0\,0\,0\,0\} \\ f(\tau) &= \{1\,2\,3\,3 \quad 3\,3\,3\,2 \quad 1\,0\,0\,0 \quad 0\,0\,0\,0\}. \end{aligned}$$

In this example we would exactly recover the ordinary convolution by cutting the circle between $\tau = 15$ and $\tau = 0$ and unrolling it. That is because the convolution sum $f(\tau)$ extends only from $\tau = 0$ to $\tau = 9$. But if a longer example were taken, for example

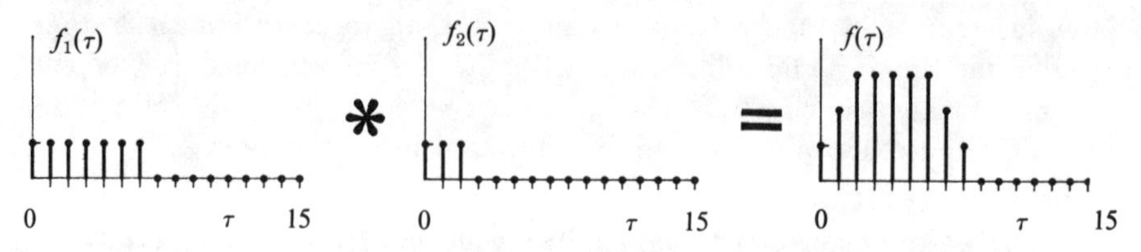

Fig. 5.1. Sequences $f_1(\tau)$ and $f_2(\tau)$ and their cyclic convolution sum $f(\tau)$.

$$
\begin{aligned}
f_1(\tau) &= \{1\,1\,1\,1 \quad 1\,1\,1\,1 \quad 1\,1\,1\,1 \quad 0\,0\,0\,0\} \\
f_2(\tau) &= \{1\,1\,1\,1 \quad 1\,1\,1\,0 \quad 0\,0\,0\,0 \quad 0\,0\,0\,0\} \\
f_{ord}(\tau) &= \{1\,2\,3\,4 \quad 5\,6\,6\,6 \quad 6\,6\,6\,6 \quad 6\,5\,4\,3 \quad 2\,1\} \\
f_{cyc}(\tau) &= \{3\,3\,3\,4 \quad 5\,6\,6\,6 \quad 6\,6\,6\,6 \quad 6\,5\,4\,3\},
\end{aligned}
$$

then $f(\tau)$ would extend from $\tau = 0$ to $\tau = 18$ as an ordinary convolution but as a cyclic convolution would complete one full period of a circle of perimeter N and overlap the locations $\tau = 0$ and $\tau = 1$. Consequently the cyclic convolution does not start {1 2 3 4 ...} as with ordinary convolution but with {3 3 3 4 ...} because the tail end of $f_{ord}(\tau)$, namely { ... 2 1}, overlaps with the start {1 2 3 4 ...}.

A program for cyclic convolution is listed in Appendix 1 under CCONV. Both CONV and CCONV implement the summations that define convolution and cyclic convolution but under some circumstances it is better to use transform methods as discussed later in the chapter. Thus the program FHTCONV performs cyclic convolution of two sequences of length 2^P by applying the convolution theorem. There is also a method using matrix multiplication, which is convenient; the speed of matrix multiplication, when implemented as a built-in machine operation, would need to be checked for each computer.

Most of what follows assumes cyclic convolution; consequently it is hardly necessary to introduce a special symbol for cyclic convolution. Nevertheless, just as a reminder, in this chapter we shall use $\circledast$. Thus

$$f_{cyc}(\tau) = f_1(\tau) \circledast f_2(\tau).$$

Packing with zeros

In some applications the cyclic convolution is precisely what is needed, overlap and all, but in other cases the ordinary convolution is desired. Then it is perfectly simple to go to larger N by packing with zeros. Both factors need to be packed to the same length and the cyclic convolution will have that length too.

Knowing where to put the zeros is not trivial. The extra zeros may lead, trail, or bracket the data but must not be inserted between adjacent data elements. If a sequence of $N-1$ elements is convolved with a sequence of N_2 elements then the number of elements in the ordinary convolution will be $N_1 + N_2 - 1$, as with $\{1\ 3\ 3\ 1\} * \{1\ 1\} = \{1\ 4\ 6\ 4\ 1\}$, where $N_1 = 4, N_2 = 2$

and the resulting convolution sequence has 5 elements. Nothing is lost if the shorter sequence is eked out with zeros to bring it into equality with the other. Thus

$$\{1\ 1\ 0\ 0\} * \{1\ 3\ 3\ 1\} = \{1\ 4\ 6\ 4\ 1\ 0\ 0\}.$$

However, if two 4-element sequences are convolved cyclically then four elements will result. Thus

$$\{1\ 1\ 0\ 0\} \circledast \{1\ 3\ 3\ 1\} = \{2\ 4\ 6\ 4\}.$$

Suppose that we wish to test a commercial software package by verifying that $\{1\ 1\ 1\ 1\} * \{1\ 1\ 1\ 1\} = \{1\ 2\ 3\ 4\ 3\ 2\ 1\}$. If we perform cyclic convolution with $N = 4$,

$$\{1\ 1\ 1\ 1\} \circledast \{1\ 1\ 1\ 1\} = \{4\ 4\ 4\ 4\}.$$

Therefore, to see the desired triangular output, extend the given sequences with zeros; then

$$\{1\ 1\ 1\ 1\quad 0\ 0\ 0\ 0\} \circledast \{1\ 1\ 1\ 1\quad 0\ 0\ 0\ 0\} = \{1\ 2\ 3\ 4\quad 3\ 2\ 1\ 0\}.$$

This is satisfactory. The output is triangular and the center of gravity has moved off to the right to $\tau = 3$ as expected. Now suppose we wish to deal with centered, or symmetrical, sequences and to verify that

$$\{1\ 1\ \underline{1}\ 1\ 1\} * \{1\ 1\ \underline{1}\ 1\ 1\} = \{1\ 2\ 3\ 4\ \underline{5}\ 4\ 3\ 2\ 1\},$$

where the underlined characters indicate the origin of τ for our purposes. Since the output has length 9 we have to extend to $N > 8$. For this example take $N = 16$ on the grounds that data analysis programs commonly provide more readily for powers of 2. Then it is true that

$$\{1\ 1\ \underline{1}\ 1\quad 1\ 0\ 0\ 0\} * \{1\ 1\ \underline{1}\ 1\quad 1\ 0\ 0\ 0\} = \{1\ 2\ 3\ 4\quad \underline{5}\ 4\ 3\ 2\quad 1\ 0\ 0\ 0\quad 0\ 0\ 0\ 0\};$$

but we can also center the packed sequences by placing the central elements at $\tau = 0$. Then

$$\{\underline{1}\ 1\ 1\ 0\quad 0\ 0\ 1\ 1\} \circledast \{\underline{1}\ 1\ 1\ 0\quad 0\ 0\ 1\ 1\} = \{\underline{5}\ 4\ 3\ 2\quad 1\ 0\ 0\ 0\quad 0\ 0\ 0\ 0\quad 1\ 2\ 3\ 4\}.$$

We see that, in the centered formulation, to get what we expect, the zeros have to be packed in at the middle rather than at the right. Failure to extend data sequences to an adequate length and to pack the extending zeros in the right place has led to much perplexity for users of other people's programs.

The short examples given above can be done by hand; an analogue calculator for facilitating cyclic convolution has two concentric dials, each with N bins in which the two sequences may be written, running in opposite senses (because of the minus sign in the definition summation). To calculate each term of the convolution, advance the rotor by one space and then sum the products of adjacent bins (Fig. 5.2). Rotary machines based on this principle were constructed in the days of electromechanical analogue computers.

Inverting convolution

It is apparent from the analogue calculator that ordinary convolution can always be inverted. Let the stator, which carries $f_2(\tau)$ on the inner of its two tracks, be furnished with an outer track on which the result $f(\tau)$ of the convolution may be written. If the result is known but the sequence $f_1(\tau)$ on the rotor is blank it would be possible to determine the missing values as follows. Positioning the blank cell for $f_1(0)$ against the given $f_2(0)$, one might say, "The first value times 1 equals 1, therefore write 1 in the first blank cell. Move that cell opposite $f_2(1)$. Then the second value times 1, plus 1 times 3, equals 5; therefore write 2 in the second blank cell." And so on.

Cyclic convolution will not be invertible, unless of course enough zeros are present to prevent overlap. Sometimes packing with zeros is an available option, sometimes not.

A simple implementation of the rotary calculator for inversion is presented in Appendix 1 under `ICONV`. Reciprocal sequences can also be determined by this method by requiring the resultant convolution to be $\{1\}$.

Convolution theorem

The convolution theorem obeyed by the DHT is as follows. If $f(\tau)$ is the cyclic convolution of $f_1(\tau)$ with $f_2(\tau)$, i.e.

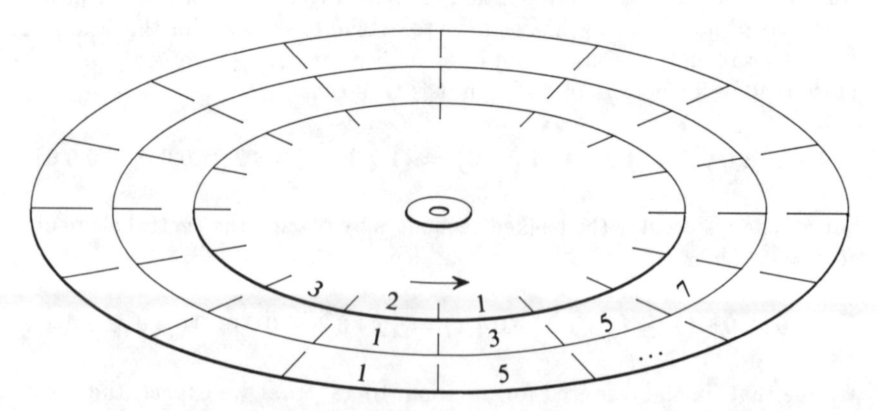

Fig. 5.2. An analogue calculator prepared for convolving $\{\underline{1}\ 2\ 3\}$ with $\{\underline{1}\ 3\ 5\ 7\}$ to give $\{\underline{1}\ 5\ 14\ 26\ 29\ 21\}$. On the rotor, $\{f_1\}$ is written backwards; $\{f_2\}$ is written on the inner track of the stator and the result $\{f_3\}$ is recorded step by step on the outer track. The rotor is in position for calculating the second term (5) of the result.

$$f(\tau) \equiv f_1(\tau) \circledast f_2(\tau) = \sum_{\tau'=0}^{N-1} f_1(\tau') f_2(\tau - \tau'),$$

then

$$H(\nu) = \tfrac{1}{2}[H_1(\nu)H_2(\nu) - H_1(-\nu)H_2(-\nu) + H_1(\nu)H_2(-\nu) + H_1(-\nu)H_2(\nu)],$$

where $H(\nu)$, $H_1(\nu)$ and $H_2(\nu)$ are the DHTs of $f(\tau)$, $f_1(\tau)$ and $f_2(\tau)$ respectively. Another way of expressing this result is in the form

$$H(\nu) = H_1(\nu)H_{2e}(\nu) + H_1(-\nu)H_{2o}(\nu),$$

where $H_2(\nu) = H_{2e}(\nu) + H_{2o}(\nu)$, the sum of its even and odd parts.

A customary way of performing convolution numerically is to take the discrete Fourier transform of each of the two given sequences and then to do complex multiplication element by element in the Fourier domain, which will require four real multiplications per element. Then one inverts the Fourier transform, remembering to change the sign of i. Analogous use of the Hartley transform for convolution will require only two real multiplications per element.

However, it very often happens, especially in image processing but also in digital filtering in general, that one of the convolving functions, say $f_2(\tau)$, is even. In that case the convolution theorem therefore simplifies to

$$H(\nu) = H_1(\nu)H_2(\nu).$$

In view of this simplified form of the theorem one need only take the DHTs of the two data sequences, multiply together term by term the two real sequences that result, and take one more DHT. Thus the proposed procedure is

$$f_1(\tau) \circledast f_2(\tau) = \text{DHT of } [(\text{DHT of } f_1) \times (\text{DHT of } f_2)].$$

This elegant and simple result resembles the corresponding result for the DFT, which differs only in that in the Fourier method one takes two direct DFTs, performs a column of *complex* multiplications, and subjects the resulting complex sequence to an *inverse* DFT. An option as to the treatment of the factor N^{-1} will be discussed below.

Inverse convolution theorem (spectral smoothing)

Sometimes convolution is to be performed in the transform domain, as for example when a spectrum $H_1(\nu)$ has to be smoothed in order to emphasize spectral features of interest. Various smoothing functions $H_2(\nu)$ have been advocated for this purpose; as the smoothing functions are always even, the applicable procedure is

$$H_1(\nu) \circledast H_2(\nu) = \text{DHT of} [(\text{DHT of} H_1) \times (\text{DHT of} H_2)].$$

Again, this is a simple and elegant result and requires only real multiplication.

Choosing the smoothing function $H_2(\nu)$ is a subjective act of judgment and it may be that the binomial coefficients $\frac{1}{4}\{2\ 1\ ...\ 1\}$, $\frac{1}{16}\{6\ 4\ 1\ ...\ 1\ 4\}$, $\frac{1}{64}\{20\ 15\ 6\ 1\ ...\ 1\ 6\ 15\}$, and so on, meet most requirements satisfactorily, reducing the choice to degree of smoothing only. Then the multiplying factor, DHT of H_2, takes the form of a simple sum of cosines; for example, DHT of $\frac{1}{4}\{2\ 1\ ...\ 1\}$ is proportional to $1 + \cos(2\pi\tau/N)$. (The constant of proportionality depends on N; however, it may be practical to retain neither this constant nor the factor $\frac{1}{4}$ during computing but to save them until the end.) The factor $1+\cos(2\pi\tau/N)$, sometimes referred to as a "cosine-squared window," is a broad weighting factor whose maximum falls at $\tau = 0$ and which falls off smoothly to zero at $\tau = N/2$. The factors corresponding to the longer binomial smoothing sequences introduce more drastic weighting factors that fall off toward zero over progressively more restricted ranges of τ.

Numerical example of convolution

A short example given above may now be used to illustrate the transform method of convolution.

In the first two columns of Table 5.1 we set out the two equalized data sequences f_1 and f_2 that are to be convolved. The next two columns show the discrete Hartley transforms H_1 and H_2. The fifth column contains the product H_1H_2 and the final column f_3 containing the convolution is N times the discrete Hartley transform of the product column. The factor N arises because of the factor N^{-1} that was adopted in the DHT definition. Instead of multiplying by N^2 throughout it might be wiser to delay the operation until normalization, calibration and other factors can be merged.

It is noticeable that the results listed under f_3 differ from the exact integral values. The differences are as large as 4 parts in 10,000 of the peak in this illustration, which was carried out with just a few decimals.

Table 5.1 Procedure for convolution

f_1	f_2	H_1	H_2	H_1H_2	f_3
2	1	0.5	2	1	6
1	4	0.427	1.457	0.622	15.008
0	6	0.25	−0.5	−0.125	20
0	4	0.073	−0.043	−0.003	15.008
0	1	0	0	0	6
0	0	0.073	0.043	0.003	0.992
0	0	0.25	−0.5	−0.125	0
1	0	0.427	−1.457	−0.622	0.992

But some approximation is in principle unavoidable in a method involving transform calculations to finite precision.

A fully general example will now be presented as Table 5.2. All of the four products arising from the convolution theorem are tabulated using the abbreviation $\bar{H}_1$ for $H(-\nu)$. The second last column is half the sum of the preceding four columns and the column headed f_3 is 8 times the DHT of the sum column. When the four product columns are examined it is clear that column 6 is the reverse of column 5 and that column 8 is the reverse of column 7. Consequently it is clearly unnecessary to perform all four multiplications; only two are needed. One way of organizing the computation is to cast the desired sum into the form

$$\tfrac{1}{2}H_1(\nu)[H_2(\nu)+H_2(-\nu)]+\tfrac{1}{2}H_1(-\nu)[H_2(\nu)-H_2(-\nu)],$$

which will be recognized as being the same as the alternative expression given in the preceding section.

By far the majority of convolutions that are performed numerically in everyday applications require only the simplified procedure illustrated in Table 5.1. In Appendix 1 a program **FHTCONV** is listed that performs this convolution using the Hartley transforms of the factors. It is best to think of this program as performing cyclic convolution but it may be used to perform ordinary convolution by extending the data with sufficient zeros to prevent overlap. It is almost always necessary to extend one of the two sequences with zeros, but if the amount of padding becomes large the advantage of the transform method will be diluted. It may be that the direct summing as in **CONV** and **CCONV** will be faster; test this possibility by comparison timing runs when the length of the shorter sequence is less than or comparable with the base-2 logarithm of the length of the longer sequence.

The factor N^{-1}

In the definition of the DHT a factor N^{-1} was included on the grounds of analogy with the DFT. But the numerical examples just given suggest

Table 5.2 Convolution of unsymmetrical sequences

f_1	f_2	H_1	H_2	H_1H_2	$-\bar{H}_1\bar{H}_2$	$H_1\bar{H}_2$	$\bar{H}_1H_2$	$\frac{1}{2}$sum	f_3
1	1	0.5	2	1	−1	1	1	1	2.992
1	2	0.552	0	0	−0.151	−0.666	0	−0.408	3.011
2	3	0	0	0	0	0	0	0	7.008
0	4	−0.125	0.207	−0.026	0	0	0.041	0.007	10.998
0	3	0.25	0	0	0	0	0	0	13.008
0	2	0.198	0	0	0.026	0.041	0	0.033	12.989
0	1	−0.25	0	0	0	0	0	0	8.992
0	0	−0.125	−1.207	0.151	0	0	−0.666	−0.258	5.002

reopening the question whether it would not have been better to use a factor $N^{-1/2}$. In that case, two applications of the same transformation in succession return the original function whereas two applications of the definition adopted here need to be followed by a further factor N to return to the original. However the apparent simplification is illusory as far as convolution is concerned because the factor N appearing in the convolution theorem does not disappear, it becomes $N^{1/2}$.

In a filtering application where the same filter coefficients are used repeatedly on different data sequences multiplication by the factor N, that was applied above after the second transformation, may be avoided by absorbing the factors into the filter coefficients.

Other applications require that the output sequence should be normalized to a sum of unity, as with probability distributions, and others again require normalization so that the initial element is unity, as with autocovariance functions. In these cases the final factor N may be absorbed into the normalizing factor.

Then again, when output sequences are displayed graphically, a scale modulus must be chosen to suit the size of the display medium. Absorbing any accumulated factors into the scale statement entirely avoids the step of modifying each output element in the course of the original computing.

For purposes of speed of computation one might as well never perform the original multiplications by N called for by the definition. When the time for normalization comes, all leftover factors can be lumped together into one calibration factor or one scale modulus.

Numerical example of autocorrelation

Examples for the autocorrelation algorithm may be given both for the simple case where the given function is even and for the general case. First consider the simple case, taking as a sample sequence {6 4 1 0 0 0 1 4}. The tabular presentation is essentially identical with that for convolution when one function is even except that now only one function is given. In Table 5.3 the third column is the DHT of the given sequence $f(\tau)$. One simply squares the entries to obtain column 4 and takes another DHT to

Table 5.3 Fast autocorrelation on an even sequence

τ, ν	$f(\tau)$	$H(\nu)$	$[H(\nu)]^2$	$C(\tau)$
0	6	2	4	70.016
1	4	1.457	2.123	56
2	1	0.5	0.125	28.032
3	0	0.043	0.002	8
4	0	0	0	1.984
5	0	0.043	0.002	8
6	1	0.5	0.25	28.032
7	4	1.457	2.123	56

get the result. As before, by working to only three decimals, we obtain approximations to the precise results, which are rational fractions.

The general case is illustrated in Table 5.4, where the given sequence is $\{1\ 2\ 3\ 4\quad 5\ 6\ 7\ 8\}$. In both of these examples one need calculate only half the values in the last column, plus one, because of symmetry.

Low pass filtering

An example of low-pass filtering was mentioned at the beginning of this chapter in connection with the smoothing of meteorological data. The procedure was to convolve with a rectangle function of width 12; consequently the result would be to produce a filter characteristic or transfer function in the form of a sinc function sinc $12f$. As it is rather helpful to understand a simple case of this kind, an example will be worked out in detail and studied. Suppose that eight years of monthly data were to be smoothed by convolution with $\frac{1}{12}\{\frac{1}{2}\ 1\ 1\ 1\ 1\ 1\ \underline{1}\ 1\ 1\ 1\ 1\ 1\ \frac{1}{2}\}$. Since there are 96 data elements and 13 filter coefficients the length of the smoothed sequence will be 108 elements. Of course in this application the first and last half dozen outputs will be valueless because an annual mean cannot be established until 12 months have elapsed. But in other applications the whole convolution is meaningful, so it will be calculated here. By selecting $N = 128$ we ensure that the sequences can be accommodated.

Our data sequence, which will be artificial, can be imagined as a monthly temperature record that contains a seasonal variation with random features superimposed on a warming trend. Increasing atmospheric carbon dioxide is threatening the pear crop that supports a rural community. As a rise in temperature of a few degrees will make pear cultivation unprofitable the trend needs study. Fig. 5.3 shows the data as square dots. A strong but irregular seasonal variation is noticeable. A warming trend is suggested by the depth of the winter minima but the summer maxima do not confirm. After smoothing with the 13 coefficients the smoothed curve which results gives a better basis for discussion.

Now if the computation is done as a cyclic convolution with $N = 128$, data values must be assigned to the 97th to 127th elements. Let these

Table 5.4 Fast autocorrelation for a general sequence

τ, ν	$f(\tau)$	$H(\nu)$	$H(-\nu)$	$\frac{1}{2}[H^2 + H(-)^2]$	$C(\tau)$
0	1	4.5	4.5	20.25	204
1	2	−1.707	0.707	1.707	175.998
2	3	−1	0	0.5	156
3	4	−0.707	−0.293	0.293	144.002
4	5	−0.5	−0.5	0.25	140
5	6	−0.293	−0.707	0.293	14.002
6	7	0	−1.0	0.5	156
7	8	0.707	−1.707	1.707	175.998

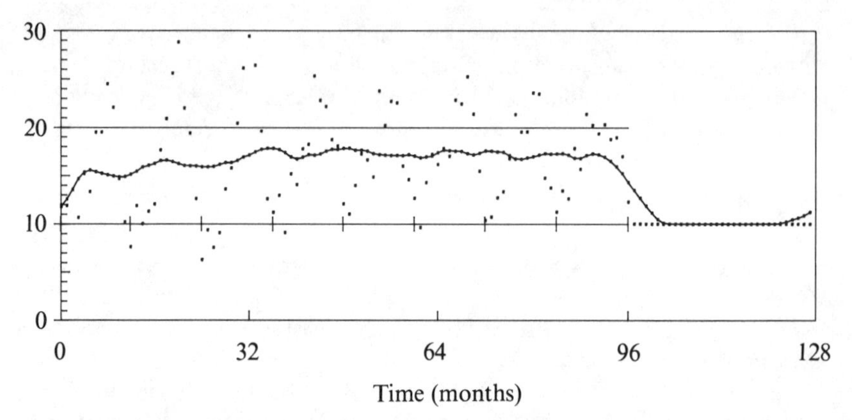

Fig. 5.3. Eight years of monthly data (square dots) and the running annual means presented at monthly intervals (curve) have been computed as a cyclic convolution with $N = 128$ and zeros have been appended to extend the data artificially to 128 values.

values be zero. A case can sometimes be made for assigning plausible values rather than implausible values; in this case one might think of using a steady value of 17, which is about the annual mean temperature. However by using zeros it will be seen very clearly from the figure where the computed cyclic convolution departs from reality. One must also fill the smoothing array with the 13 desired coefficients and 115 zeros. For the figure $\frac{1}{12}\{1\ 1\ 1\ 1\ 1\ 1\ \frac{1}{2}\ 0\ ...\ 0\ \frac{1}{2}\ 1\ 1\ 1\ 1\ 1\}$ was used for consistency. Thus each smoothed value agrees in time with the middle of the 13 values that entered into the mean. If one used $\frac{1}{12}\{\frac{1}{2}\ 1\ 1\ 1\ 1\ 1\ 1\ 1\ 1\ 1\ 1\ 1\ \frac{1}{2}\ 0\ 0\ 0\ ...\ 0\ 0\}$ then there would be a difference of indexing – the smoothed sequence would be delayed six months. The effect of cyclic convolution has been to produce 128 output values, only 84 of which (those from months 7 to 90) are meaningful. These values are preceded and followed by more-or-less linearly tapered ramps; a small piece of the rising ramp appears at the extreme right. The range where the output is zero clearly reflects the artificial data used to extend the array to 128 values and is not likely to be misinterpreted. In other circumstances there may be no guard zone of zeros and the rising and falling ramps would then overlap, perhaps giving an appearance of plausible output. One way to avoid misinterpretation is to determine which output values are significant and to plot only those. The smoothed output presented in Fig. 5.3 would be the same whether the cyclic convolution was computed in the data domain by multiplications and additions or whether a transform method was used.

Edge enhancement

Much of the information in an image resides in the boundaries between more or less homogeneous regions and the eye appeal of an image may be improved by enhancing such boundaries. One such type of boundary is the edge between regions of different average brightness. To improve the contrast in an aerial photograph between a cultivated field of brightness 7 and a surrounding pasture of brightness 8 one might subtract 6 throughout. The field would then have brightness 1 and the pasture would be twice as bright with brightness 2 or, with amplification, the levels would be 4 and 8 and the contrast would be greatly enhanced. Such adjustments can be done in the photographic dark room on analogue images or by computation on digital images. However there may be other areas of the complete image where the brightness does not permit such subtraction. Edge enhancement aims at emphasizing edges regardless of the absolute level of brightness in the vicinity of the edge. If an edge is characterized by a steep rate of change of brightness then one could locate steep slopes and make them steeper; to do this one would simply subtract a suitable amount of the second derivative in a direction normal to the edge. An attractive feature of this idea is that the subtraction could be uniformly applied all over without any decision-making as to where the edges are.

In one dimension, convolution with $\{1 \quad -2 \quad 1\}$ takes the second difference, which will be an appropriate representation of second derivative provided the sampling interval is fine enough. If it is not, it is unreasonable to ask for what is in essence improved resolution; on the other hand, if the sampling is very fine relative to the resolution interval of interest, as is usually the case, some longer sequence, such as samples of the second derivative of a Gaussian, would be appropriate.

In two dimensions the local curvature is estimated correspondingly, for example by convolution with an array of samples of $(\partial^2/\partial r^2)\exp(-\pi r^2/W^2)$, where the scale modulus W is chosen to be smaller than the interval of interest but not small compared with the sampling interval. For illustration consider the following small array of only 9 elements

$$\begin{bmatrix} 1 & 2 & 1 \\ 2 & -12 & 2 \\ 1 & 2 & 1 \end{bmatrix}.$$

Convolving two dimensionally with the image to be enhanced, and subtracting a fraction of the result from the image, will have the desired effect. The amount to be subtracted will involve a quality judgment that will vary from case to case and cannot be decided on mathematical grounds alone. The complete enhancing operation can be compressed into one stage; for example, if 1/16 of the "curvature" map is to be subtracted, the final enhanced result is obtainable directly by convolution with

$$\frac{1}{16}\begin{bmatrix} -1 & -2 & -1 \\ -2 & 28 & -2 \\ -1 & -2 & -1 \end{bmatrix}.$$

Convolution by FHT

The procedure for convolution presented in Table 5.1 may be implemented using the fast algorithm as in the program **FHTCONV** of Appendix 1. The main program assumes that one or other of the given functions possesses symmetry because in that case, which is common, the program runs very fast. However, if symmetry is absent, a simple subroutine numbered 4400 is supplied as a substitute for subroutine 4000, which simply multiplies the two real DHTs. Use of the fast algorithm **FHTSUB** imposes lengths that are multiples of two, a familiar restriction that is not particularly onerous. Of more concern is the fact that the data sets to be convolved are not at all likely in general to be of the same length, whereas the program requires both sequences to be extended to the same length with zeros. If the data sets are not of comparable size it may well be faster to convolve in the data domain using **CONV**.

When the result is printed a divisor N is included, or $2N$ in the unsymmetrical case, following a strategy described above in the section on the factor N^{-1}. There are three visits to the fast subroutine 9000 in this program; consequently it would be a waste of time to repeat the pretabulation of trigonometric functions and powers of two. The subroutine itself recognizes this fact and adapts automatically.

Where cyclic convolution is not wanted, a convolution sum of length N means that only $\frac{1}{2}N$ data values should be provided for each input function. Thus, in addition to any zeros provided in order to extend the length to a multiple of two, a further string of $\frac{1}{2}N$ zeros is required.

Convolution by multiplication is excellent in the right application, easy to implement when the housekeeping details relating to zeros are attended to, and much cleaner than the corresponding complex version.

Dispensing with permutation

Convolution can be performed entirely without permutation at the price of storing a second type of transform. The algorithms discussed in this book make use of a permutation stage followed by nested loops, but there is a second kind of algorithm in which the permutation stage follows the nested loops. For a description of this structure with reference to the DFT see W.T. Cochran *et al.*, "What is the Fast Fourier Transform?", *IEEE Trans. Audio and Electroacoustics*, vol. AU-15, pp. 45-55, 1985.

In terms of notation introduced in Chapter 7, the algorithm that prepermutes applies a matrix operator $\mathbf{L}_P\mathbf{L}_{P-1}...\mathbf{L}_2\mathbf{L}_1\mathbf{P}_N$; the algorithm that postpermutes applies $\mathbf{P}_N\mathbf{L}_1\mathbf{L}_2...\mathbf{L}_{P-1}\mathbf{L}_P$.

If the second kind of algorithm were used on the two input functions, and the first kind were used on the product of the outputs to return to the original domain, clearly there would be no need to permute. Fast permutation, as described later, means that the time saving is a small fraction of the whole, and the fraction diminishes with N. However, there may be specialized circumstances where avoidance of permutation makes sense.

Convolution as matrix multiplication

The convolution sum of two sequences is expressible as the matrix product of a certain rectangular matrix on[1] a column matrix. The rules for matrix multiplication require that the width M of the rectangle equal the height N of the column; the resultant column matrix will have height $M + N - 1$. Thus the following convolution between a five-element and a three-element sequence

$$\{f(0) \quad f(1) \quad f(2) \quad f(3) \quad f(4)\} * \{g(0) \quad g(1) \quad g(2)\}$$

is equivalent to

$$\begin{bmatrix} f(0) & 0 & 0 \\ f(1) & f(0) & 0 \\ f(2) & f(1) & f(0) \\ f(3) & f(2) & f(1) \\ f(4) & f(3) & f(2) \\ 0 & f(4) & f(3) \\ 0 & 0 & f(4) \end{bmatrix} \times \begin{bmatrix} g(0) \\ g(1) \\ g(2) \end{bmatrix} = \begin{bmatrix} f(0)g(0) \\ f(1)g(0) + f(0)g(1) \\ f(2)g(0) + f(1)g(1) + f(0)g(2) \\ f(3)g(0) + f(2)g(1) + f(1)g(2) \\ f(4)g(0) + f(3)g(1) + f(2)g(2) \\ f(4)g(1) + f(3)g(2) \\ f(4)g(2) \end{bmatrix}.$$

Since convolution is a commutative operation, it follows that there is an alternative form

$$f * g = \begin{bmatrix} g(0) & 0 & 0 \\ g(1) & g(0) & 0 \\ g(2) & g(1) & g(0) \\ g(3) & g(2) & g(1) \\ g(4) & g(3) & g(2) \\ 0 & g(4) & g(3) \\ 0 & 0 & g(4) \end{bmatrix} \times \begin{bmatrix} f(0) \\ f(1) \\ f(2) \end{bmatrix},$$

which, from the point of view of matrix theory, is nonobvious. On computers that provide for matrix multiplication it is not necessary to have a special program for convolution.

Appendix 1 lists such a program under MATCON. Array $B(\ )$ is filled with N data values (lines 70-90) as in the case of CONV. Then the first column of array $A(,)$, which is to be two-dimensional, is filled with the M data values and extended with zeros to total length $P = M + N - 1$ (line 120) and the remaining columns of $A(,)$ are filled to build up the circulant structure

[1] Use of the transitive preposition 'on' rather than the conjunctive preposition 'with' in the phrase 'product of **A** on **B**,' helps remind us of the noncommutative property of matrix multiplication.

(lines 130-150). With this preparation the whole convolution is carried out and assigned to $C(\ )$ by the one-line matrix multiplication statement

```
180 MAT C = A*B.
```

Problems

5.1 *Autocorrelation exercise.* Find the cyclic autocorrelation of the following sequences of length $N = 8$.
(a) $\{1\,1\,1\,1 \quad 0\,0\,0\,0\}$,
(b) $\{1\,1\,1\,1 \quad 1\,0\,0\,0\}$,
(c) $\{1\,1\,1\,1 \quad 1\,1\,0\,0\}$,
(d) $\{1\,1\,1\,1 \quad 1\,1\,1\,0\}$.

5.2 *Autocorrelation exercise.* What is the cyclic autocorrelation of
(a) $\{1\,0\,1\,0 \quad 1\,0\,1\,0\}$,
(b) $\{1\,0\,0\,1 \quad 0\,0\,1\,0\}$,
(c) $\{1\,0\,0\,0 \quad 1\,0\,0\,1\}$?
(d) Sum the elements of the autocorrelation sequences and explain the results.

5.3 *Autocorrelation exercise.* Find the cyclic autocorrelation of
(a) $\{1\,1\,0\,0 \quad 0\,0\,0\,0\}$,
(b) $\{1\,0\,1\,0 \quad 0\,0\,0\,0\}$,
(c) $\{1\,1\,0\,1 \quad 0\,0\,0\,0\}$,
(d) $\{1\,1\,0\,0 \quad 1\,0\,1\,0\}$.

5.4 *Autocorrelation exercise.* Find the cyclic autocorrelation of
(a) $\{-3\ -2\ -1\ 0 \quad 1\,2\,3\,4\}$,
(b) $\{-4\ -3\ -2\ -1 \quad 1\,2\,3\,4\}$.
(c) Sum the elements of the autocorrelation sequences and explain the result.

5.5 *Autocorrelation sequences.* Find the autocorrelation of
(a) $\{1\,1\,0\,0 \quad 1\,0\,1\,0 \quad 0\,0\,0\,0 \quad 0\,0\,0\,0\}$,
(b) $\{0\,0\,0\,0 \quad 0\,0\,0\,0 \quad 1\,1\,0\,0 \quad 1\,0\,1\,0\}$.
(c) Evaluate the product $1,100,101 \times 1,010,011$ and use the result to deduce an extremely simple autocorrelation algorithm. (d) If the method you have arrived at is so simple, why is it not better known?

5.6 *Inverse autocorrelation.* Experiment with `ICORR` (Appendix 1) to invert autocorrelation sequences established in the previous problems. Obtain the sequence whose autocorrelation is $\{1.1\ 6\ 15\ 20\ 15\ 6\ 1.1\}$.

5.7 *Convolution exercise.* Convolve $\{1\,1\,1\,1 \quad 0\,0\,0\,0\}$ with
(a) $\{1\,1\,0\,0 \quad 0\,0\,0\,0\}$,
(b) $\{1\,1\,1\,0 \quad 0\,0\,0\,0\}$,
(c) $\{1\,1\,1\,1 \quad 1\,0\,0\,0\}$,
(d) $\{1\,1\,1\,1 \quad 1\,1\,1\,1\}$.

5.8 *Convolution exercise.* Convolve $\{1\ \frac{1}{2}\ \frac{1}{4}\ \frac{1}{8} \quad \frac{1}{16}\ \frac{1}{32}\ \frac{1}{64}\ \frac{1}{128}\}$ with
(a) $\{1\ -1\ 0\ 0 \quad 0\,0\,0\,0\}$, (b) $\{2\ -1\ 0\ 0 \quad 0\ 0\ 0\ -1\}$.

5.9 *Convolution with odd factor.* A data sequence is to be convolved with an odd sequence, i.e. one such that $f(N - \tau) = -f(\tau)$. After the DHTs are taken only N products are required. Show that reversal of the transform of the *data* sequence before multiplication and retransformation will lead to the desired result.

5.10 *Compaction.* A data sequence of 1024 elements has to be compacted to 256 at the sacrifice of fine detail. Show that this can be done by taking the DHT with $N = 1024$ and then taking the DHT of the first quarter of the transform elements using $N = 256$.

5.11 *Running means.* (a) Calculate the running mean of five consecutive terms of the binomial sequence $\{0\,0\,0\,0 \quad 1\,4\,6\,4 \quad 1\,0\,0\,0\}$,
(b) What is the variance of the resulting sequence?

5.12 *Cyclic sinc function.* Show that $\operatorname{sinc} x+\operatorname{sinc}(x-N)+\operatorname{sinc}(x-2N)+ \ldots + \operatorname{sinc}(x+N)+\operatorname{sinc}(x+2N)+\operatorname{sinc}(x+3N)\ldots = N^{-1}\sin(N\pi x)/\sin(\pi x)$.

CHAPTER 6

TWO-DIMENSIONAL TRANSFORMS

"One picture is worth a thousand words."

Emperor Sung

Images on surfaces have always been of importance and are becoming more so as technical means for creating, modifying and presenting such images are developed. Digital telemetry of planetary images from spacecraft, and digital processing of these images, have become familiar as a result of space exploration and the design of analogue optical imaging systems has made remarkable advances. Analogue optical image processing is a reality and optical digital processing of two-dimensional digital data is in view. In all these fields spectral analysis is a customary tool; consequently there are corresponding applications for the Hartley transform.

The two-dimensional Hartley transform proves to have an interesting fundamental property. We have become accustomed to the notion that the Hartley transform is real in one dimension, and so it is in two dimensions. Because the Fourier transform is complex, half the transform plane suffices to determine the Fourier transform. The rest of the plane is occupied by conjugate values that bear no additional information, because points that are diametrically opposite are labeled with conjugate coefficients. In the Hartley plane, by contrast, there is no such symmetry and no redundancy; the information is spread half as thick over the whole area. This distinction, in different embodiments for different applications, may favor the choice of one transform over the other. We now turn to the definition of the two-dimensional Hartley transforms, continuous and discrete.

The two-dimensional Hartley transform

Starting from a function $f(x, y)$ we define its two-dimensional Hartley transform $H(u, v)$, and deduce the inverse transform, to obtain:

$$H(u,v) = \int_{-\infty}^{\infty}\int_{-\infty}^{\infty} f(x,y)\,\mathrm{cas}[2\pi(ux+vy)]\,dx\,dy$$

$$f(x,y) = \int_{-\infty}^{\infty}\int_{-\infty}^{\infty} H(u,v)\,\mathrm{cas}[2\pi(ux+vy)]\,du\,dv.$$

In terms of the Fourier transform $F(u, v) = R(u, v) + iI(u, v)$ it follows that $H(u, v) = R(u, v) - I(u, v)$ as may be verified from the Fourier relations that are presented here for comparison.

$$F(u,v) = \int_{-\infty}^{\infty}\int_{-\infty}^{\infty} f(x,y)e^{-i2\pi(ux+vy)}dxdy$$

$$f(x,y) = \int_{-\infty}^{\infty}\int_{-\infty}^{\infty} F(u,v)e^{i2\pi(ux+vy)}dudv.$$

The inverse transform may be established by taking the Hartley transform of $H(\ ,\ )$ and showing that the result is the original function $f(\ ,\ )$. The reasoning depends on symmetry properties.

Symmetry and antisymmetry

A given function $f(x,y)$ may be decomposed into symmetrical and antisymmetrical parts such that

$$f(x,y) = f_{symm}(x,y) + f_{antisymm}(x,y),$$

where

$$f_{symm}(x,y) = \tfrac{1}{2}[f(x,y) + f(-x,-y)]$$

and

$$f_{antisymm}(x,y) = \tfrac{1}{2}[f(x,y) - f(-x,-y)].$$

This resolution into parts is the two-dimensional generalization of splitting functions of one variable into even and odd parts. Just as the real part of the Fourier transform is even and the imaginary part is odd, so the real part of the two-dimensional Fourier transform is symmetrical and the imaginary part is antisymmetrical. It follows that the real two-dimensional Hartley transform expressed as $R(u,v) - I(u,v)$ is already presented in terms of its symmetrical and antisymmetrical parts.

To establish the inverse transform quoted above we now take the two-dimensional Hartley transform of the transform $H(u,v)$. Let

$$\hat{f}(x,y) = \text{HT of } H(u,v) = \text{HT of } R(u,v) - \text{HT of } I(u,v).$$

Now HT of $R(u,v)$ = FT of $R(u,v) = f_{symm}(x,y)$ because $R(u,v)$ is symmetrical. Likewise HT of $I(u,v) = -f_{antisymm}(x,y)$. So

$$\hat{f}(x,y) = f_{symm}(x,y) - [-f_{antisymm}(x,y)] = f(x,y), \text{ Q.E.D.}$$

Examples

There is no particular difficulty in evaluating two-dimensional Hartley transforms and a few examples will serve as models. In Table 6.1, $\mathbf{H}(x)$

Table 6.1 Some two-dimensional Hartley transform pairs

$f(x,y)$	$H(u,v)$
$e^{-\pi[(x-a)^2+(y-b)^2]}$	$e^{-\pi(u^2+v^2)}\,\mathrm{cas}[2\pi(au+bv)]$
$\Pi(x-\frac{1}{2})\,\Pi(y-\frac{1}{2})$	$\mathrm{sinc}\,u\,\mathrm{sinc}\,v\,\mathrm{cas}[\pi(u+v)]$
$\delta(x-a)\delta(y-b)$	$\mathrm{cas}[2\pi(au+bv)]$
$\delta(x-a)$	$\mathrm{cas}\,2\pi au\;\delta(v)$
$\delta(x\cos\theta+y\sin\theta-a)$	$\mathrm{cas}[2\pi a(u\cos\theta+v\sin\theta)]\delta(v\sin\theta-u\cos\theta)$
$e^{-x}\mathrm{H}(x)$	$(1+2\pi u)(1+4\pi^2u^2)^{-1}\delta(v)$
$e^{-(x\cos\theta+y\sin\theta)}\mathrm{H}(x\cos\theta+y\sin\theta)$	$\dfrac{1+2\pi(u\cos\theta+v\sin\theta)}{1+4\pi^2(u\cos\theta+v\sin\theta)^2}$
$\sin\pi x$	$-\frac{1}{2}\delta(u+\frac{1}{2})\delta(v)+\frac{1}{2}\delta(u-\frac{1}{2})\delta(v)$
$\cos\pi x$	$\frac{1}{2}\delta(u+\frac{1}{2})\delta(v)+\frac{1}{2}\delta(u+\frac{1}{2})\delta(v)$
$(x-x^2)\,\Pi(x-\frac{1}{2})\,\Pi(y)$	$\frac{1}{2}(\pi^2u^2)^{-1}\,\mathrm{cas}\,\pi u[\mathrm{sinc}\,u-\cos\pi u]\,\mathrm{sinc}\,v$
$\sin\pi x\,\Pi(x-\frac{1}{2})\,\Pi(y)$	$\frac{1}{2}\,\mathrm{cas}\,\pi u\;[\mathrm{sinc}(u+\frac{1}{2})+\mathrm{sinc}(u-\frac{1}{2})]\,\mathrm{sinc}\,v$
$\Pi[\sqrt{(x-a)^2+y^2}]$	$\frac{1}{2}\,\mathrm{cas}\,2\pi au\;(u^2+v^2)^{-1}J_1(\pi\sqrt{u^2+v^2})$

is the Heaviside unit step function and $\Pi(x)$ is the unit rectangle function $\mathrm{H}(x+\frac{1}{2})-\mathrm{H}(x-\frac{1}{2})$.

Theorems in two dimensions

Table 6.2 presents the generalizations of the well-known theorems to two dimensions, both for the Fourier and Hartley transforms so that the similarities and differences may be easily seen. Table 6.3 shows those theorems that are expressible in the form of relations between properties in the two domains.

Circular symmetry

When $f(x,y)$ is circularly symmetrical and may be represented by, let us say, $\mathbf{f}(r)$, then the two-dimensional transform is the same as the two-dimensional Fourier transform because the Fourier transform in this case has no imaginary part. Consequently the two-dimensional Hartley transform reduces to the one-dimensional Hankel transform $\mathcal{H}(q)$ of $\mathbf{f}(r)$:

$$\int_{-\infty}^{\infty}\int_{-\infty}^{\infty} f(x,y)\,\mathrm{cas}[2\pi(ux+vy)]\,dx dy = \int_0^{\infty}\mathbf{f}(r)J_0(2\pi qr)2\pi r\,dr = \mathcal{H}(q).$$

In this formulation r is the radial variable in the (x,y)-plane and

$$q=\sqrt{u^2+v^2}$$

Table 6.2 Theorems for two-dimensional Fourier and Hartley transforms

Theorem	$f(x,y)$	$F(u,v)$	$H(u,v)$
Similarity	$f(x/a, y/b)$	$\lvert ab\rvert\, F(au+bv)$	$\lvert ab\rvert\, H(au+bv)$
Addition	$f_1(x,y)+f_2(x,y)$	$F_1(u,v)+F_2(u,v)$	$H_1(u,v)+H_2(u,v)$
Shift	$f(x-a, y-b)$	$e^{-i2\pi(au+bv)}F(u,v)$	$\cos 2\pi(au+bv)\, H(u,v)$ $+\sin 2\pi(au+bv)\, H(-u,-v)$
Modulation	$f(x,y)\times$ $\cos 2\pi(u_0x+v_0y)$	$\frac{1}{2}F(u-u_0, v-v_0)$ $+\frac{1}{2}F(u+u_0, v+v_0)$	$\frac{1}{2}H(u-u_0, v-v_0)$ $+\frac{1}{2}H(u+u_0, v+v_0)$
Convolution	$f_1(x,y) ** f_2(x,y)$	$NF_1(u,v)F_2(u,v)$	$\frac{1}{2}N[H_1H_2 - H_1(\text{-,-})H_2(\text{-,-})$ $+H_1H_2(\text{-,-}) + H_1(\text{-,-})H_2]$
Autocorrelation	$f(x,y) \star\star f(x,y)$	$N\lvert F(u,v)\rvert^2$	$N[H(u,v)]^2 + N[H(-u,-v)]^2$
Separable product	$f_1(x)f_2(y)$	$F_1(u)F_2(v)$	$\frac{1}{2}[H_1H_2 + H_1(\text{-})H_2$ $+H_1H_2(\text{-}) - H_1(\text{-})H_2(\text{-})]$
Rotation	$f(x', y')$	$F(u', v')$	$H(u', v')$

Table 6.3 Theorems for relations between domains

Theorem	Property	Fourier relation	Hartley relation
Infinite integral	$\int\int f\,dxdy$	$=F(0,0)$	$=H(0,0)$
Rayleigh's	$\int\int f^2\,dxdy$	$=\int\int FF^*(u,v)\,dudv$	$=\int\int HH(\text{-,-})\,dudv$
First moment	$\int\int xf\,dxdy$	$=\frac{1}{-i2\pi}\frac{\partial F}{\partial u}\Big\rvert_{0,0}$	$=\frac{1}{2\pi}\frac{\partial H}{\partial u}\Big\rvert_{0,0}$
Second moment	$\int\int x^2 f\,dxdy$	$=\frac{1}{(-i2\pi)^2}\frac{\partial^2 F}{\partial u^2}\Big\rvert_{0,0}$	$=\frac{1}{(2\pi)^2}\frac{\partial^2 H}{\partial u^2}\Big\rvert_{0,0}$
Bessel inequality	$\lvert f\rvert$	$\le\int\int \lvert F\rvert\,dudv$	$\le\int\int\sqrt{\lvert HH(\text{-,-})\rvert}\,dudv$

is the radial variable in the (u, v)-plane. The one-dimensional Hartley transform of $\mathcal{H}(q)$ is the Abel transform $\mathcal{A}(x)$ of $\mathbf{f}(r)$, where, by definition,

$$\mathcal{A}(x) = 2\int_x^\infty \frac{\mathbf{f}(r)r\,dr}{\sqrt{r^2-x^2}}.$$

Thus

$$\int_{-\infty}^{\infty} \mathcal{H}(q) \operatorname{cas} 2\pi x q \, dq = \mathcal{A}(x)$$

and conversely.

There are no features arising in connection with circular symmetry that distinguish the Hartley from the Fourier transform. Consequently any operations ordinarily carried out on circularly symmetrical functions by means of Fourier transforms may be carried out instead by Hartley transforms and the results will be the same.

Filtering in two dimensions

Manipulation of two-dimensional digital images also benefits from the existence of a real transform. An image $f(\tau_1, \tau_2)$ represented by an $N_1 \times N_2$ matrix possesses a two-dimensional discrete Hartley transform (^{2}DHT) which is itself an $N_1 \times N_2$ matrix $H(\nu_1, \nu_2)$ of real numbers. The transformation and its inverse are

$$H(\nu_1, \nu_2) = \frac{1}{N_1 N_2} \sum_{\tau_1=0}^{N_1-1} \sum_{\tau_2=0}^{N_2-1} f(\tau_1, \tau_2) \operatorname{cas}\left(\frac{2\pi\nu_1\tau_1}{N_1} + \frac{2\pi\nu_2\tau_2}{N_2}\right)$$

$$f(\tau_1, \tau_2) = \sum_{\nu_1=0}^{N_1-1} \sum_{\nu_2=0}^{N_2-1} H(\nu_1, \nu_2) \operatorname{cas}\left(\frac{2\pi\nu_1\tau_1}{N_1} + \frac{2\pi\nu_2\tau_2}{N_2}\right).$$

A two-dimensional spatial frequency (ν_1, ν_2) describes an obliquely oriented cas function which has ν_1/N_1 cycles per unit of τ_1 in the east-west direction and ν_2/N_2 cycles per unit of τ_2 in the north-south direction.

Properties of the two-dimensional discrete transform, including the convolution theorem, which is the basis for digital filtering of images, are deducible from tables given in the next chapter. For the most part the relation between the continuous and discrete is the same for two dimensions as for one. However, the interpretation of cyclicity introduced earlier in this chapter deserves special mention.

In one dimension the variable τ ranges from 0 to $N-1$ and the domain of τ can be pictured as a straight row of N equispaced points. But the domain can also be pictured as a ring of N points with the advantage that values of τ that are negative or greater than $N-1$ can be given a meaning. The value to be assigned to any integral τ is then expressible by

$$f(\tau) = f(\tau \bmod N).$$

In two dimensions one can adopt the same plan and write

$$f(\tau_1, \tau_2) = f(\tau_1 \bmod N_1, \tau_2 \bmod N_2).$$

Thus with $N_1 = 32$ and $N_2 = 16$ the point $(\tau_1, \tau_2) = (34, -5)$ does not lie on the flat rectangular domain of $N_1 N_2$ points arranged in N_1 columns and N_2 rows. But a value can still be assigned to $f(34, -5)$. Noting that 34 mod 32 is 2 and -5 mod 16 is 11, we assign the value $f(2, 11)$. This is just a matter of arithmetic that can be taken care of in a computer but for thinking about the consequences it is desirable to have in mind the spatial topology corresponding to the ring arrangement on the plane. The usual prescription is to roll the rectangular domain into a cylinder so as to bring the top and bottom rows adjacent to each other. Then bend the cylinder into a toroid by bringing the left and right end rings into adjacency. Thus in Fig. 6.1 *A* is brought into contact with *E* and *B* with *F* and the circular ends *ACE* and *BDF* of the resulting cylinder are brought into contact by bending the cylinder. The bending step is the harder to picture because something has to give and the originally square cells defined by the points cannot continue to have equal sides. In toroidal chess the players do not bother with a toroidal chess board; they know that the white bishop (Fig. 6.1) can move along the diagonal marked by an arrow and after passing off the right hand edge of the board can continue on as shown by the longer arrow. It is rather a dangerous game. The white rook threatens the black rook and knights always have eight possible moves regardless of their position on the board; thus the white knight, whose moves are marked by dots, can capture the black rook. This is the best way to visualize the space of τ_1 and τ_2; it is as though the nuclear region bounded by (0,0) and (7,7) is replicated ad infinitum in both directions.

Cyclicity in two dimensions

Generalizing the discrete transform to two dimensions does not introduce any surprises except perhaps for the properties of cyclicity. Just as in one dimension we agree that $f(\tau \pm N) = f(\tau)$ so in two dimensions

$$f(\tau_1 \pm N, \tau_2 \pm N) = f(\tau_1, \tau_2).$$

This does not present any difficulty algebraically and the topology may be clearly visualized by appeal to the toroidal surface which is the generaliza-

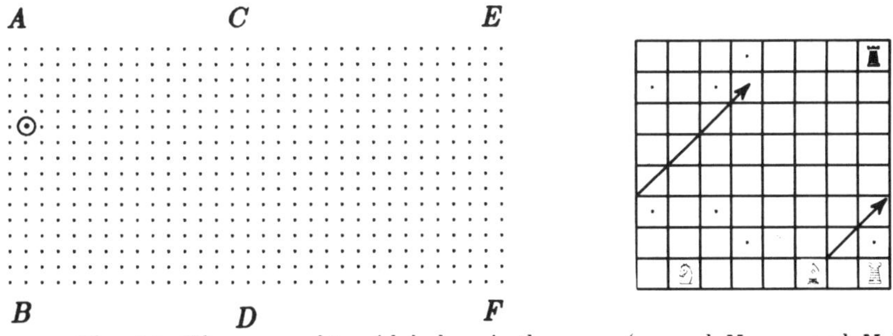

Fig. 6.1. The space of toroidal chess is the space (τ_1 mod N_1, τ_2 mod N_2). The point $(34, -5)$ on the 32×16 array is the same as the point $(2, 11)$ shown as $\odot$.

tion of the ring in one dimension. There is a possible source of confusion, however, in the indexing scheme, which can be a real trap for beginners.

Suppose that a digital object has to be smoothed by convolution with the following array of coefficients

$$\frac{1}{256}\begin{bmatrix} 1 & 4 & 6 & 4 & 1 \\ 4 & 16 & 24 & 16 & 4 \\ 6 & 24 & 36 & 24 & 6 \\ 4 & 16 & 24 & 16 & 4 \\ 1 & 4 & 6 & 4 & 1 \end{bmatrix}.$$

The factor 1/256 ensures that the low spatial frequencies are not amplified. It is understood that the central value 36 will be placed on a pixel of choice, that the coefficients will be multiplied by the elements on which they fall, that the products will be summed and that the resulting value will be written in the chosen pixel. When it comes to representing this simple operation by two-dimensional convolution with a digital object the central value 36 is indexed at (1,1). Consequently, the $N \times N$ array representing the smoothing function is (leaving empty space where there are only zeros)

$$\frac{1}{256}\begin{bmatrix} 36 & 24 & 6 & & & 6 & 24 \\ 24 & 16 & 4 & & & 4 & 16 \\ 6 & 4 & 1 & & & 1 & 4 \\ & & & & & & \\ & & & & & & \\ 6 & 4 & 1 & & & 1 & 4 \\ 24 & 16 & 4 & & & 4 & 16 \end{bmatrix}.$$

This organization seems so unsymmetrical and bizarre to people who naturally feel that the array should be centered, that their subconscious minds rebel. Yet to proceed otherwise would be to shift the image.

A second practical consideration that may get mixed up with cyclicity arises when the convolving array does not possess symmetry. For example, suppose that an imaging system possesses astigmatism such that a point object produces an asymmetric response

$$\begin{bmatrix} 1 & 5 & 10 & 6 & 1 \\ 7 & 14 & 25 & 20 & 5 \\ 5 & 18 & 36 & 30 & 10 \\ 2 & 15 & 20 & 16 & 4 \\ 1 & 4 & 6 & 3 & 1 \end{bmatrix}$$

whose centroid is evidently offset towards the upper right. With what array must an extended object be convolved in order to obtain an image exhibiting the effects of this astigmatism? To answer this question, first rotate the array by 180° in its own plane to take account of the two minus signs that are present in the definition of convolution in two dimensions. Then set the nominal central value 36 at (1,1) as before. The array to be entered into convolution is thus

$$\begin{bmatrix} 36 & 18 & 5 & & 10 & 30 \\ 25 & 14 & 7 & & 5 & 20 \\ 10 & 5 & 1 & & 1 & 6 \\ & & & & & \\ & & & & & \\ 6 & 4 & 1 & & 1 & 3 \\ 20 & 15 & 2 & & 4 & 16 \end{bmatrix}.$$

Three dimensions

The transform formulae generalize regularly to three (and more) dimensions and have applications to crystallography, plasma physics, gas dynamics and many other fields. The three-dimensional relations will simply be given for reference without further discussion.

$$H(\nu_1,\nu_2,\nu_3) = \frac{1}{N_1 N_2 N_3} \sum_{\tau_1=0}^{N_1-1} \sum_{\tau_2=0}^{N_2-1} \sum_{\tau_3=0}^{N_3-1} f(\tau_1,\tau_2,\tau_3) \times \operatorname{cas}\left(\frac{2\pi\nu_1\tau_1}{N_1} + \frac{2\pi\nu_2\tau_2}{N_2} + \frac{2\pi\nu_3\tau_3}{N_3}\right),$$

$$f(\tau_1,\tau_2,\tau_3) = \sum_{\nu_1=0}^{N_1-1} \sum_{\nu_2=0}^{N_2-1} \sum_{\nu_3=0}^{N_3-1} H(\nu_1,\nu_2,\nu_3) \times \operatorname{cas}\left(\frac{2\pi\nu_1\tau_1}{N_1} + \frac{2\pi\nu_2\tau_2}{N_2} + \frac{2\pi\nu_3\tau_3}{N_3}\right).$$

Problems

6.1 *Numerical examples.* Obtain the discrete two-dimensional Hartley transforms of the following small digital images. Take the origin $(\tau_1, \tau_2) = (0,0)$ to be in the top left-hand corner as is usual with a matrix.

$$\begin{matrix} 1 & 1 & 1 & 1 \\ 1 & 1 & 1 & 1 \\ 1 & 1 & 1 & 1 \\ 1 & 1 & 1 & 1 \end{matrix} \qquad \begin{matrix} 1 & 2 & 1 & 0 \\ 2 & 4 & 2 & 0 \\ 1 & 2 & 1 & 0 \\ 0 & 0 & 0 & 0 \end{matrix} \qquad \begin{matrix} 1 & 3 & 3 & 1 \\ 3 & 6 & 6 & 3 \\ 3 & 6 & 6 & 3 \\ 1 & 3 & 3 & 1 \end{matrix}$$

$$\begin{matrix} 0 & 0 & 0 & 0 \\ 0 & 1 & 1 & 0 \\ 0 & 1 & 1 & 0 \\ 0 & 0 & 0 & 0 \end{matrix} \qquad \begin{matrix} 1 & 0 & 0 & 1 \\ 0 & 1 & 1 & 0 \\ 0 & 1 & 1 & 0 \\ 1 & 0 & 0 & 1 \end{matrix} \qquad \begin{matrix} 1 & 2 & 3 & 4 \\ 5 & 6 & 7 & 8 \\ 9 & 10 & 11 & 12 \\ 13 & 14 & 15 & 16 \end{matrix}$$

6.2 *Point spread function.* The transform of a digital object is operated upon by multiplication with the following various two-dimensional low-pass filter characteristics. If the object is a point object, what does the image become in each case?

$$\begin{matrix} 1 & 1 & 0 & 1 \\ 1 & 0 & 0 & 0 \\ 0 & 0 & 0 & 0 \\ 1 & 0 & 0 & 0 \end{matrix} \qquad \begin{matrix} 1 & 0 & 0 & 0 \\ 0 & 1 & 0 & 1 \\ 0 & 0 & 0 & 0 \\ 0 & 1 & 0 & 1 \end{matrix} \qquad \begin{matrix} 1 & 1 & 0 & 1 \\ 1 & 1 & 0 & 1 \\ 0 & 0 & 0 & 0 \\ 1 & 1 & 0 & 1 \end{matrix} \qquad \begin{matrix} 2 & 1 & 0 & 1 \\ 1 & 0 & 0 & 0 \\ 0 & 0 & 0 & 0 \\ 1 & 0 & 0 & 0 \end{matrix}$$

6.3 *Shift theorem.* Derive the shift theorem in two dimensions.

6.4 *Separable product.* Derive the separable product theorem.

6.5 *Modulation theorem.* Explain why the modulation theorem is exactly the same for the HT as for the FT.

6.6 *Quonset hut.* Show that the Hartley transform of $\sqrt{a^2 - x^2}\,\Pi(x/2a)\,\Pi(y/4a)$ is $(a^2/2x)J_1(2\pi a x)\,\mathrm{sinc}(4av)$.

6.7 *Shifting an even function.* If $f(x, y)$ is an even function with respect to both x and y, show that $f(x - a, y - b)$ has HT $\mathrm{cas}[2\pi(au + bv)]H(u, v)$.

6.8 *Transform pairs.* Verify the entries in Table 6.1.

6.9 *Elementary transforms.* Obtain the discrete Hartley transforms of the following single-element functions and explain the character of each transform.

$$\begin{matrix} 0 & 1 & 0 & 0 \\ 0 & 0 & 0 & 0 \\ 0 & 0 & 0 & 0 \\ 0 & 0 & 0 & 0 \end{matrix} \qquad \begin{matrix} 0 & 0 & 1 & 0 \\ 0 & 0 & 0 & 0 \\ 0 & 0 & 0 & 0 \\ 0 & 0 & 0 & 0 \end{matrix} \qquad \begin{matrix} 0 & 0 & 0 & 1 \\ 0 & 0 & 0 & 0 \\ 0 & 0 & 0 & 0 \\ 0 & 0 & 0 & 0 \end{matrix}$$

$$\begin{matrix} 0 & 0 & 0 & 0 \\ 0 & 1 & 0 & 0 \\ 0 & 0 & 0 & 0 \\ 0 & 0 & 0 & 0 \end{matrix} \qquad \begin{matrix} 0 & 0 & 0 & 0 \\ 0 & 0 & 0 & 0 \\ 0 & 0 & 0 & 0 \\ 0 & 0 & 0 & 1 \end{matrix} \qquad \begin{matrix} 0 & 0 & 0 & 0 \\ 0 & 0 & 0 & 0 \\ 0 & 0 & 1 & 0 \\ 0 & 0 & 0 & 0 \end{matrix}$$

6.10 *Packing with zeros.* Given that the discrete Hartley transform of $\begin{bmatrix} a & b \\ c & d \end{bmatrix}$ is $\begin{bmatrix} \alpha & \beta \\ \gamma & \delta \end{bmatrix}$, find the DHT of

$$\begin{bmatrix} \alpha & 0 & \beta & 0 \\ 0 & 0 & 0 & 0 \\ \gamma & 0 & \delta & 0 \\ 0 & 0 & 0 & 0 \end{bmatrix}.$$

6.11 *Diffraction pattern of slit.* With a view to gaining a hint as to the diffraction pattern of a slit, and perhaps later of other apertures, contemplate the transforms of two large square digital images: one with 1s down the left-hand column and zeros elsewhere, the other with 1s on the main diagonal and zeros elsewhere. (a) Work out the transforms for $N_1 = N_2 = 4$ and deduce the results for larger images. (b) Is the DHT the same as the DFT? (c) Do the results agree with what the rotation theorem for continuous variables would suggest?

6.12 *Choice of origin.* The origin of a digital image is taken in the bottom left-hand corner as for cartesian coordinates rather than in the top left-hand corner as for matrices. A programmer says that it will make no difference to the Hartley transform except that, of course, the transform will be upside down. Is that true?

6.13 *Four-element coefficients.* (a) Tabulate the four 4-element patterns $\mathrm{cas}[2\pi(\nu_1\tau_1/2 + \nu_2\tau_2/2)]$ in the τ-domain for all ν_1 and ν_2, making sure that the row and column conventions are correct. (b) Hence show that

$$\begin{bmatrix} a & b \\ c & d \end{bmatrix} \text{ has 2D DHT } \tfrac{1}{4}\begin{bmatrix} a+b+c+d & a-b+c-d \\ a+b-c-d & a-b-c+d \end{bmatrix}.$$

6.14 *Eight-element coefficients.* Construct the sixteen 16-element patterns $\mathrm{cas}[2\pi(\nu_1\tau_1/4 + \nu_2\tau_2/4)]$ in the τ-domain.

6.15 *Special transform.* A new transform has been proposed that is defined by

$$Q(u,v) = \int_{-\infty}^{\infty}\int_{-\infty}^{\infty} f(x,y)\,\mathrm{cas}(2\pi ux)\,\mathrm{cas}(2\pi vy)\,dx\,dy.$$

Show that if this transformation is applied twice in succession to a real function that the original function is regained. Try out a few cases numerically with a view to discovering how this special transform is related to the Hartley transform.

CHAPTER 7

FACTORIZATION OF THE TRANSFORM MATRIX

"And thou shalt set in it settings of stones, even four rows of stones: the first row shall be a sardius, a topaz, and a carbuncle: this shall be the first row. And the second row shall be an emerald, a sapphire, and a diamond. And the third row a ligure, an agate, and an amethyst. And the fourth row a beryl, and an onyx, and a jasper: they shall be set in gold in their inclosings."

Exodus 28: 15

One way of looking at the summation that defines the discrete Hartley transform is as a square matrix operating on an N-dimensional vector. This point of view will be developed in what follows because it leads to an understanding of the basis for the fast algorithm; each factor of the matrix corresponds to one stage, or subroutine, of the algorithm. The matrix view illuminates the step of permutation, where the matrix involved can be looked at as simply a graph, or scatter diagram, of the output versus the input indices. The step from the Hartley transform to the Fourier transform can also be expressed as a matrix multiplication and so the factorization of the Hartley matrix leads directly to a new factorization of the Fourier matrix; this is the feature that gives importance to the present chapter.

Matrix formulation of discrete operator

No extensive familiarity with matrix algebra is required to read this material as the only operation used is the product of a square matrix on a column matrix. In that case the result is another column matrix whose rth element is the sum of the product of the elements of the rth row of the square matrix with the elements of the given column matrix. Thus

$$\begin{bmatrix} a & b & c & d \\ e & f & g & h \\ i & j & k & l \\ m & n & o & p \end{bmatrix} \begin{bmatrix} A \\ B \\ C \\ D \end{bmatrix} = \begin{bmatrix} aA+bB+cC+dD \\ eA+fB+gC+hD \\ iA+jB+kC+lD \\ mA+nB+oC+pD \end{bmatrix}.$$

If one were to expand the summation specified in the discrete transform formulae there would be N separate equations. Thus, to take the modest case of $N=4$, the DFT equations corresponding to

$$F(\nu) = N^{-1} \sum_{\tau=0}^{N-1} f(\tau) e^{-i2\pi\nu\tau/N}$$

would be

$$F(0) = \tfrac{1}{4}[f(0) + f(1) + f(2) + f(3)]$$
$$F(1) = \tfrac{1}{4}[f(0) + f(1)e^{-i2\pi/N} + f(2)e^{-i2\pi 2/N} + f(3)e^{-i2\pi 3/N}]$$

$$F(2) = \tfrac{1}{4}[f(0) + f(1)e^{-i2\pi 2/N} + f(2)e^{-i2\pi 4/N} + f(3)e^{-i2\pi 6/N}]$$
$$F(3) = \tfrac{1}{4}[f(0) + f(1)e^{-i2\pi 3/N} + f(2)e^{-i2\pi 6/N} + f(3)e^{-i2\pi 9/N}].$$

Such a set of equations lends itself to expression in terms of matrix multiplication relating an input 4-vector

$$\{f(0) \quad f(1) \quad f(2) \quad f(3)\}$$

to an output vector

$$\{F(0) \quad F(1) \quad F(2) \quad F(3)\}.$$

A nice way of compacting the matrix equation below is to use the abbreviation

$$W = e^{-i2\pi/N}$$

since the exponential factor $\exp(-i2\pi/N)$ appears in most of the terms; and where it does not appear it may be supposed that the factor is W^0, which is unity.

In matrix notation the equations reduce to a single equation

$$\begin{bmatrix} F(0) \\ F(1) \\ F(2) \\ F(3) \end{bmatrix} = \begin{bmatrix} 1 & 1 & 1 & 1 \\ 1 & W & W^2 & W^3 \\ 1 & W^2 & W^4 & W^6 \\ 1 & W^3 & W^6 & W^9 \end{bmatrix} \begin{bmatrix} f(0) \\ f(1) \\ f(2) \\ f(3) \end{bmatrix}.$$

The square matrix represents the operator that converts the sequence $f(\cdot)$ into the Fourier transform sequence $F(\cdot)$. Using bold letters to represent matrices we can write (omitting N^{-1} in the spirit of p. 45)

$$\mathbf{F} = \mathbf{Wf},$$

where $\mathbf{f}$ is the column matrix whose elements are $f(0)$, $f(1)$, ... $f(N-1)$, $\mathbf{F}$ is the column matrix whose elements are $F(0)$, $F(1)$, ... $F(N-1)$ and

$$\mathbf{W} = \begin{bmatrix} 1 & 1 & 1 & & & & \\ 1 & W & W^2 & & & & \\ 1 & W^2 & W^4 & & & & \\ & & & \cdot & & & \\ & & & & \cdot & & \\ & & & & W^{(N-3)^2} & W^{(N-3)(N-2)} & W^{(N-1)(N-3)} \\ & & & & W^{(N-2)(N-3)} & W^{(N-2)^2} & W^{(N-2)(N-1)} \\ & & & & W^{(N-1)(N-3)} & W^{(N-1)(N-2)} & W^{(N-1)^2} \end{bmatrix}.$$

Similarly there is another matrix $\mathbf{\Psi}$ which operates on the input sequence $\mathbf{f}$ to produce the discrete Hartley transform sequence $\mathbf{H}$:

$$\mathbf{H} = \mathbf{\Psi f}.$$

An interesting thing about the matrix $\mathbf{W}$ is that it can be factorized in a way that leads to the Fast Fourier Transform algorithm (FFT). A factorization of the matrix $\mathbf{\Psi}$ in the form

$$\mathbf{\Psi} = \mathbf{L}_P \mathbf{L}_{P-1} ... \mathbf{L}_3 \mathbf{L}_2 \mathbf{L}_1 \mathbf{P}_N \mathbf{f}$$

will now be presented that leads to a new and distinctive fast algorithm for computing the discrete Hartley transform matrix $\mathbf{H}$.

The smallest value of N that allows the nature of the factorization to be made clear is 16, in which case there are five matrix factors and a constant factor N^{-1}. If $N = 2^P$, then there are $P + 1$ matrix factors.

The first matrix factor, the one that operates first on $\mathbf{f}$, is a permutation operator $\mathbf{P}_N$. The P subsequent operators $\mathbf{L}_s$ may be termed stage matrices. If $N = 16$ then $P = 4$ and there are 4 stages; if $N = 32$ there are 5 stages, and so on.

Permutation

Permutation is a rearrangement of the order of the elements of the input matrix $\mathbf{f}$ that may succinctly be described in terms of a "perfect shuffle." A deck of cards may be shuffled by cutting into two equal halves and riffling them together. If each card from one half lodges between cards of the other half it is a perfect shuffle (Fig. 7.1). The same terminology may be applied to sequences: we say the transition

$$\{a \quad b \quad c \quad d \quad e \quad f \quad g \quad h\} \quad \text{to} \quad \{a \quad e \quad b \quad f \quad c \quad g \quad d \quad h\}$$

is a perfect shuffle. The inverse transition

$$\{a \quad b \quad c \quad d \quad e \quad f \quad g \quad h\} \quad \text{to} \quad \{a \quad c \quad e \quad g \quad b \quad d \quad f \quad h\}$$

is an inverse shuffle.

Permutation of a sequence of length N consists of first making an in-

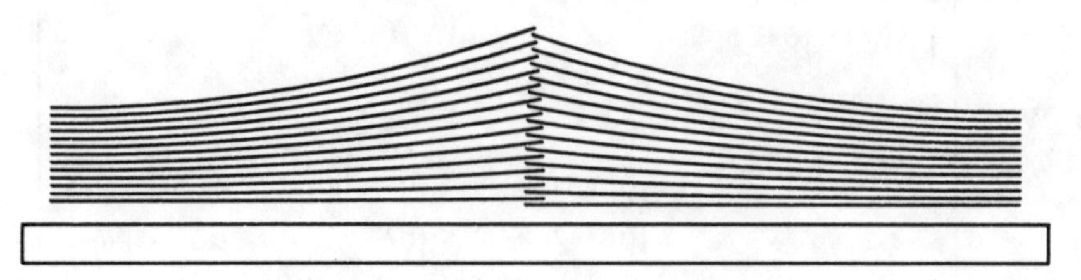

Fig. 7.1. The perfect shuffle.

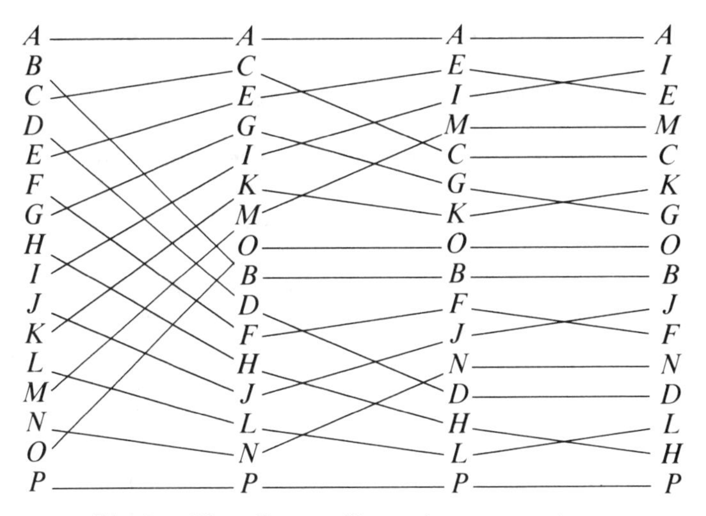

Fig. 7.2. Flow diagram illustrating permutation.

verse shuffle on the whole sequence. Then each half is given an inverse shuffle. Then each quarter is given an inverse shuffle and so on until the subdivision produces groups of four, and when the fours are inversely shuffled that terminates the process. For a detailed view of the permutation of a 16-element sequence we exhibit the 3 successive rearrangements in rows below each other.

Initial sequence: $\{a\ b\ c\ d\ e\ f\ g\ h\ i\ j\ k\ l\ m\ n\ o\ p\}$,
First shuffle: $\{a\ c\ e\ g\ i\ k\ m\ o\ b\ d\ f\ h\ j\ l\ n\ p\}$,
Second shuffle: $\{a\ e\ i\ m\ c\ g\ k\ o\ b\ f\ j\ n\ d\ h\ l\ p\}$,
Third shuffle: $\{a\ i\ e\ m\ c\ k\ g\ o\ b\ j\ f\ n\ d\ l\ h\ p\}$.

In this case, where $P = 4$, the third perfect shuffle produces the desired permutation. Another way of presenting the process is by a flow diagram as in Fig. 7.2.

Permutation diagrams

Suppose that the ith element in a sequence of length $N = 2^P$ becomes the jth element upon permutation. Then j is a function of i and N that may be written

$$j = P_N(i).$$

Fig. 7.3 displays the character of this function for various values of N. The graphs are reminiscent of statistical scatter diagrams with no apparent

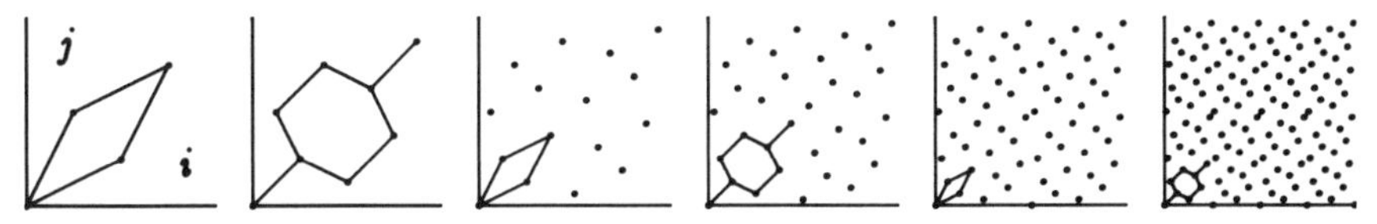

Fig. 7.3. Permutation diagrams of $j = P_N(i)$ for increasing values of N. Each diagram is reduced in scale by a factor 2 relative to the one on its left.

systematic dependance of j upon i. In fact the correlation coefficient is zero in each case but of course the relation between j and i is perfectly deterministic. In these diagrams j is plotted upwards as is usual with cartesian coordinates. However, the successive rows of matrices are placed below each other as in European writing; consequently the permutation diagrams are upside down relative to the permutation matrices presented below. For reference, Table 7.1 presents a number of permuted sequences.

Permutation matrix

A permutation matrix $\mathbf{P}_N$ can now be defined to comprise an $N \times N$ array with 1s at $(i, j) = [i, P_N(i)]$ and 0s elsewhere. Thus

$$\mathbf{P}_8 = \begin{bmatrix} 1 & & & & & & & \\ & & & & 1 & & & \\ & & 1 & & & & & \\ & & & & & & 1 & \\ & 1 & & & & & & \\ & & & & & 1 & & \\ & & & 1 & & & & \\ & & & & & & & 1 \end{bmatrix}$$

and

$$\mathbf{P}_{16} = \begin{bmatrix} 1 & & & & & & & & & & & & & & & \\ & & & & & & & & 1 & & & & & & & \\ & & & & 1 & & & & & & & & & & & \\ & & & & & & & & & & & & 1 & & & \\ & & 1 & & & & & & & & & & & & & \\ & & & & & & & & & & 1 & & & & & \\ & & & & & & 1 & & & & & & & & & \\ & & & & & & & & & & & & & & 1 & \\ & 1 & & & & & & & & & & & & & & \\ & & & & & & & & & 1 & & & & & & \\ & & & & & 1 & & & & & & & & & & \\ & & & & & & & & & & & & & 1 & & \\ & & & 1 & & & & & & & & & & & & \\ & & & & & & & & & & & 1 & & & & \\ & & & & & & & 1 & & & & & & & & \\ & & & & & & & & & & & & & & & 1 \end{bmatrix}.$$

Cell structure of permutation

If the diagrams of Fig. 7.3 are studied, or Table 7.1, a repetitive character will be discovered which is illustrated in Fig. 7.4. On subdividing

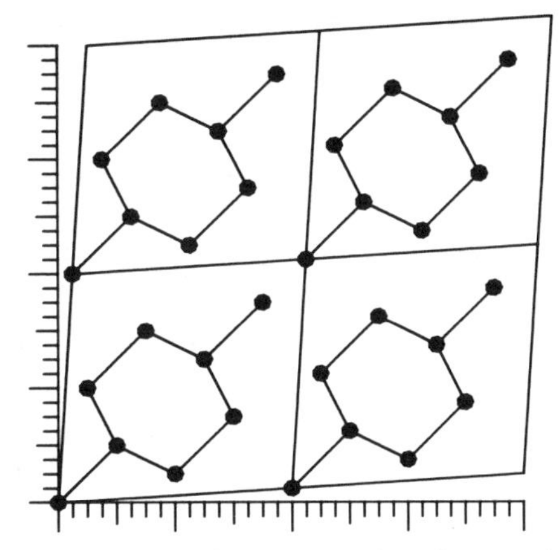

Fig. 7.4. With $N = 32$, the permutation diagram is like a crystal with four cells and eight atoms per cell.

the permutation diagram for $N = 32$ into four rhomboidal cells, and connecting the eight atoms within each cell to emphasize the pattern, one sees a crystal structure repeating on a nonrectangular basis in both directions. Interestingly, the content of each cell is geometrically similar to the permutation diagram for $N = 8$, but doubled in scale. Now if the whole $N = 32$ diagram is doubled in scale it will be found to be a repeating component of the $N = 128$ diagram.

The diagrams for an even index P ($N = 8$, 32, 128, ...) form a family but the diagrams for odd P ($N = 4$, 16, 64, ...) form a second family. This topology is not at all obvious from the defining prescription as summarized in Fig. 7.2 but is noticeable when numbers of permutation matrices are presented for inspection. It is apparent that there are two different kinds of cell; family I has cells with four atoms arranged like a kite while family II has cells with eight atoms arranged like a frog. It is easy to verify the following characteristics.

Parameter	Family I	Family II
Index, P	even	odd
Atoms per cell	4	8
Number of cells, C	$N/4$	$N/8$
Grid spacing in cell, N_0	$\sqrt{N/4}$	$\sqrt{N/8}$
Cells per side, N_0	$\sqrt{N/4}$	$\sqrt{N/8}$
Cell size, N/N_0	$\sqrt{4N}$	$\sqrt{8N}$

*Table 7.1. Permuted sequences for various values of **N***

$N = 8$		$N = 16$		$N = 32$		$N = 64$		$N = 64$		$N = 128$	
i	j	i	j	i	j	i	j	i	j	i	j
0	0	0	0	0	0	0	0	36	9	0	0
	4		8		16		32		41		64
2	2		4		8		16		25		32
	6		12		24		48		57		96
4	1						8				16
	5	4	2	4	4		40	40	5		80
6	3		10		20		24		37		48
	7		6		12		56		21		112
			14		28				53		
						8	4		13	8	8
		8	1	8	2		36		45		72
			9		18		20		29		40
			5		10		52		61		104
			13		26		12				24
							44	48	3		88
		12	3	12	6		28		35		56
			11		22		60		19		120
			7		14				51		
			15		30	16	2		11	16	4
							34		43		68
				16	1		18		27		36
					17		50		59		100
					9		10				20
					25		42	56	7		84
							26		39		52
					5		58		23		116
					21				55		
					13	24	6		15	24	12
					29		38		47		76
							22		31		44
				24	3		54		63		108
					19		14				28
					11		46				92
					27		30				60
							62				124
					7						
					23	32	1			32	2
					15		33				66
					31		17				34
							49				98

$N = 2$	
i	j
0	0
1	1

$N = 4$	
i	j
0	0
1	2
2	1
3	3

Table 7.1, contd.

$N = 128$		$N = 128$		$N = 128$		$N = 256$		$N = 256$	
i	j	i	j	i	j	i	j	i	j
	18	72	9	104	11	0	0	32	4
	82		73		75		128		132
	50		41		43		64		68
	114		105		107		192		196
			25		27		32		36
40	10		89		91		160		164
	74		57		59		96		100
	42		121		123		224		228
	106								
	26	80	5	112	7	8	16	40	20
	90		69		71		144		148
	58		37		39		80		84
	122		101		103		208		212
			21		23		48		52
48	6		85		87		176		180
	70		53		55		112		116
	38		117		119		240		244
	102								
	22	88	13	120	15	16	8	48	12
	86		77		79		136		140
	54		45		47		72		76
	118		109		111		200		204
			29		31		40		44
56	14		93		95		168		172
	78		61		63		104		108
	46		125		127		232		236
	110						88		
	30	96	3			24	24	56	28
	94		67				152		156
	62		35				88		92
	126		99				216		220
			19				56		60
64	1		83				184		188
	65		51				120		124
	33		115				248		252
	97								
	17								
	81								
	49								
	113								

Although the two figures make it clear that there is a cell pattern that repeats, the structure that results when each cell is replaced by a single point is not as regular and must be investigated. In the permutation diagram for $N = 128$, mark the lower left element of each cell. One easily verifies that the coordinates of the μth cell in the νth row for $N = 128$ are

$$x_{\mu,\nu} = 4\mu + P_4(\nu)$$

$$y_{\mu,\nu} = 4\nu + P_4(\mu).$$

The factor 4 arises from the fact that $x_{\mu+1,\nu} - x_{\mu,\nu} = 4$, the grid spacing of the atoms within one cell. The subscript 4 represents the number of cells per side. Since the total number of atoms is N and there are 4 or 8 atoms per cell, the total number of cells is $C = N/(4 \text{ or } 8)$ and the number of cells per side, N_0, is the square root. Thus

$$N_0 = \sqrt{\frac{N}{4 \text{ or } 8}} = \sqrt{\frac{N}{4(1 + P \bmod 2)}}.$$

The grid spacing within a cell is the cell width N/N_0 divided by the 4 or 8 atoms contained and is thus equal to N_0, which is why the number 4 appears twice in a cell coordinate equation. In general

$$x_{\mu,\nu} = N_0\mu + P_{N_0}(\nu)$$

$$y_{\mu,\nu} = N_0\nu + P_{N_0}(\mu).$$

The topological structure described here is the basis for a fast permutation algorithm **FASTPERMUTE** described later.

Slow permutation, which is traditional, is performed as follows. To find j, given i and N, convert i to binary notation using P bits, write it down backwards, and reconvert to decimal. For example, with $N = 16$ we know that $P_{16}(7) = 14$. The binary representation of $i = 7$ is 0111 and 1110 is the binary representation of 14.

A handy way of permuting by bit reversal when brevity seems more desirable than speed is as follows:

```
4000 D$ = DTB$(I)

4010 J = BTD(REV$(D$[17 - P, 16]))
```

The function **DTB$()** converts its decimal argument into a binary string of 16 characters, **REV$()** reverses its string argument, and **BTD()** converts its binary string argument back to decimal form.

Stage matrices

Following permutation comes a succession of P operations leading stage by stage to the final transform. The stage number, which will be called s, ranges from 1 to P. The general form of the operators $\mathbf{L}_s$ is best conveyed in the case of $N = 16$ by beginning with the last operator $\mathbf{L}_4$. We adopt the following abbreviations for cosines of submultiples of 2π.

$$S_{n,s} = \sin 2\pi n/2^s, \quad C_{n,s} = \cos 2\pi n/2^s, \quad K_{n,s} = \operatorname{cas} 2\pi n/2^s.$$

With these conventions the 4th stage matrix $\mathbf{L}_4$ is given by

$\mathbf{L}_4 =$

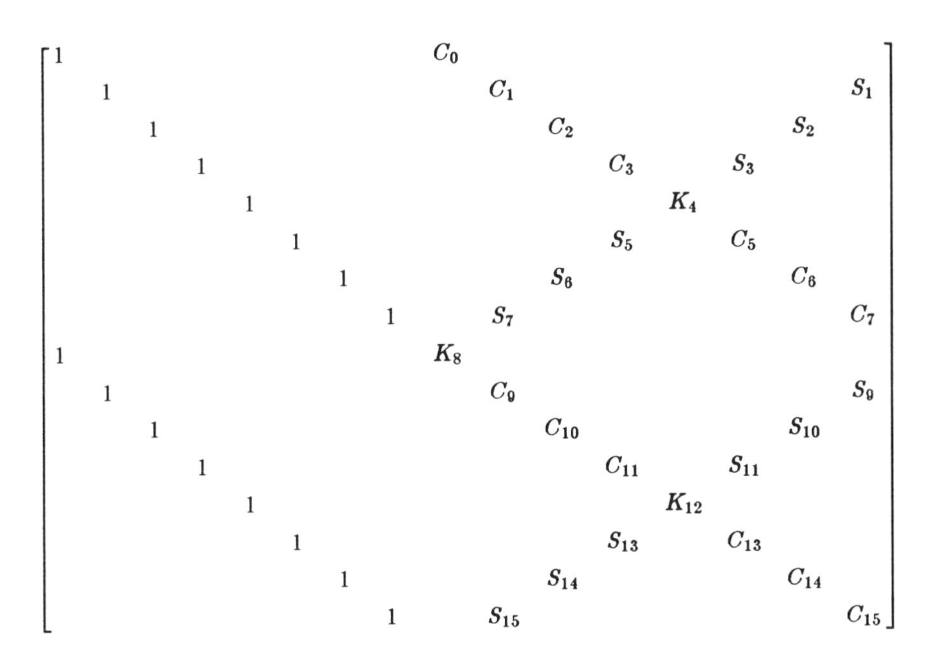

$$\begin{bmatrix}
1 & & & & & & & & C_0 & & & & & & & \\
 & 1 & & & & & & & & C_1 & & & & & & S_1 \\
 & & 1 & & & & & & & & C_2 & & & & S_2 & \\
 & & & 1 & & & & & & & & C_3 & & S_3 & & \\
 & & & & 1 & & & & & & & & K_4 & & & \\
 & & & & & 1 & & & & & & S_5 & & C_5 & & \\
 & & & & & & 1 & & & & S_6 & & & & C_6 & \\
 & & & & & & & 1 & & S_7 & & & & & & C_7 \\
1 & & & & & & & & K_8 & & & & & & & \\
 & 1 & & & & & & & & C_9 & & & & & & S_9 \\
 & & 1 & & & & & & & & C_{10} & & & & S_{10} & \\
 & & & 1 & & & & & & & & C_{11} & & S_{11} & & \\
 & & & & 1 & & & & & & & & K_{12} & & & \\
 & & & & & 1 & & & & & & S_{13} & & C_{13} & & \\
 & & & & & & 1 & & & & S_{14} & & & & C_{14} & \\
 & & & & & & & 1 & & S_{15} & & & & & & C_{15}
\end{bmatrix}$$

In this matrix, where $s = 4$ throughout, the subscript s is omitted so that C_n stands for $C_{n,4} = \cos 2\pi n/16$ and so on. There are three nonzero elements in each row, which means that three elements of the operand column matrix enter, with appropriate coefficients, into each output element.

The cosine elements and the matrix elements that are equal to one are distributed on lines parallel to the main diagonal but the sine elements run the other way. As a result, the independent variable of the operand element that multiplies the sine coefficient S_n diminishes as n increases. This property is referred to as retrograde indexing.

Now $\mathbf{L}_3$ divides into two null quarters and two quarters that are the same as $\mathbf{L}_4$ but smaller. Nevertheless the retrograde indexing can be seen:

$$
\mathbf{L}_3 = \begin{bmatrix}
1 & & & & C_0 \\
& 1 & & & & C_1 & & S_1 \\
& & 1 & & & & K_2 \\
& & & 1 & & S_3 & & C_3 \\
1 & & & & C_4 \\
& 1 & & & & C_5 & & S_5 \\
& & 1 & & & & K_4 \\
& & & 1 & & S_6 & & C_6 \\
& & & & & & & & 1 & & & & C_0 \\
& & & & & & & & & 1 & & & & C_1 & & S_1 \\
& & & & & & & & & & 1 & & & & K_2 \\
& & & & & & & & & & & 1 & & S_3 & & C_3 \\
& & & & & & & & 1 & & & & C_4 \\
& & & & & & & & & 1 & & & & C_5 & & S_5 \\
& & & & & & & & & & 1 & & & & K_6 \\
& & & & & & & & & & & 1 & & S_7 & & C_7
\end{bmatrix}
$$

In the matrix $\mathbf{L}_3$, since the subscript 3 is omitted, C_n stands for $C_{n,3} = \cos 2\pi n/8$.

The earlier stages $\mathbf{L}_2$ and $\mathbf{L}_1$ follow the same pattern.

$$
\mathbf{L}_2 = \begin{bmatrix}
1 & & 1 \\
& 1 & & 1 \\
1 & & -1 \\
& 1 & & -1 \\
& & & & 1 & & 1 \\
& & & & & 1 & & 1 \\
& & & & 1 & & -1 \\
& & & & & 1 & & -1 \\
& & & & & & & & 1 & & 1 \\
& & & & & & & & & 1 & & 1 \\
& & & & & & & & 1 & & -1 \\
& & & & & & & & & 1 & & -1 \\
& & & & & & & & & & & & 1 & & 1 \\
& & & & & & & & & & & & & 1 & & 1 \\
& & & & & & & & & & & & 1 & & -1 \\
& & & & & & & & & & & & & 1 & & -1
\end{bmatrix}
$$

$$\mathbf{L}_1 = \begin{bmatrix} 1 & 1 & & & & & & & & & & & & & & \\ 1 & -1 & & & & & & & & & & & & & & \\ & & 1 & 1 & & & & & & & & & & & & \\ & & 1 & -1 & & & & & & & & & & & & \\ & & & & 1 & 1 & & & & & & & & & & \\ & & & & 1 & -1 & & & & & & & & & & \\ & & & & & & 1 & 1 & & & & & & & & \\ & & & & & & 1 & -1 & & & & & & & & \\ & & & & & & & & 1 & 1 & & & & & & \\ & & & & & & & & 1 & -1 & & & & & & \\ & & & & & & & & & & 1 & 1 & & & & \\ & & & & & & & & & & 1 & -1 & & & & \\ & & & & & & & & & & & & 1 & 1 & & \\ & & & & & & & & & & & & 1 & -1 & & \\ & & & & & & & & & & & & & & 1 & 1 \\ & & & & & & & & & & & & & & 1 & -1 \end{bmatrix}$$

Since $C_{n,2} = \cos 2\pi n/4$ and $S_{n,2} = \sin 2\pi n/4$ can assume only values -1, 0 or 1, the general pattern is not conspicuous in $\mathbf{L}_2$ although the matrix is consistent with the pattern, and the same is true of $\mathbf{L}_1$.

To show how the matrices develop as N increases we move up to $N = 32$. Then $\mathbf{L}_4$ is composed of two null quarters and two quarters identical with $\mathbf{L}_4$, as it was for $N = 16$. The new stage matrix $\mathbf{L}_5$ expands regularly as would be expected.

To summarize for $N = 16$, the DHT is derived from the input sequence $\mathbf{f}$ by

$$\begin{aligned} \mathbf{H} &= \mathbf{\Psi f} \\ &= \mathbf{L}_4\mathbf{L}_3\mathbf{L}_2\mathbf{L}_1\mathbf{P}_{16}\mathbf{f}, \end{aligned}$$

where the discrete Hartley matrix $\mathbf{\Psi}$, reduced to the factors presented, is

$$\mathbf{\Psi} = \mathbf{L}_4\mathbf{L}_3\mathbf{L}_2\mathbf{L}_1\mathbf{P}_{16}\mathbf{f}.$$

In general, for any $N = 2^P$

$$\mathbf{\Psi} = \mathbf{L}_P\mathbf{L}_{P-1} \ldots \mathbf{L}_2\mathbf{L}_1\mathbf{P}_N\mathbf{f}.$$

Conversion to DFT

Finally, to convert the DHT to the DFT is a simple matter of associating the even and odd parts of the DHT with the real and imaginary parts of the DFT. The matrix operator that creates twice the even part is

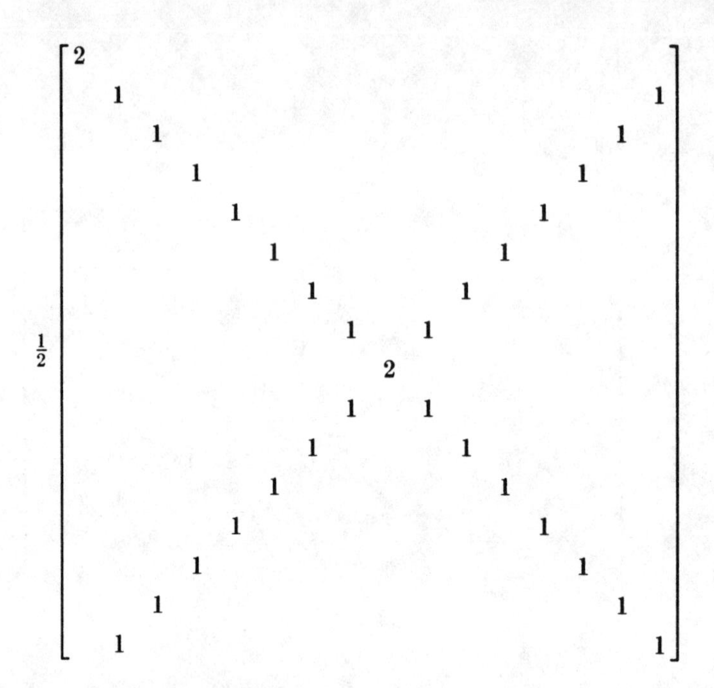

while twice the odd part is created by

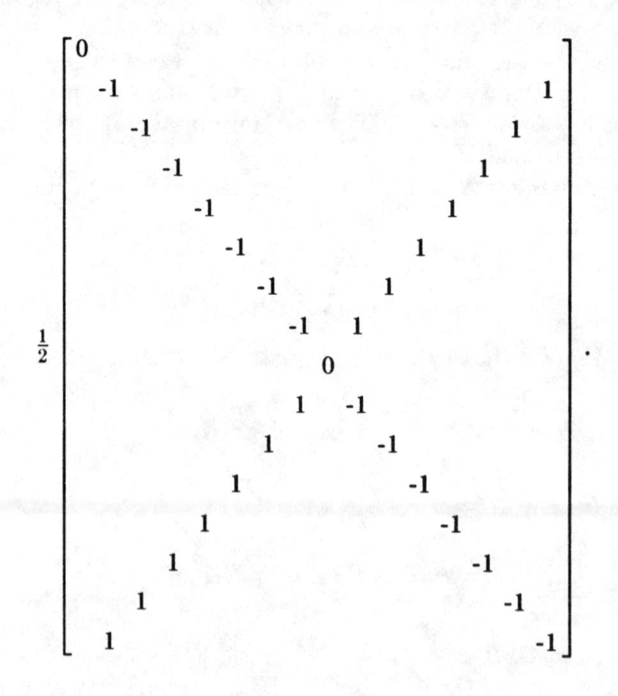

Multiplying the second by $-i$ and adding to the first gives the conversion matrix $\mathbf{\Phi}$ that generates the DFT:

$2\mathbf{\Phi} =$

$$\begin{bmatrix}
2 &&&&&&&&&&&&&&& \\
& 1-i &&&&&&&&&&&&&& 1+i \\
&& 1-i &&&&&&&&&&&& 1+i & \\
&&& 1-i &&&&&&&&&& 1+i && \\
&&&& 1-i &&&&&&&& 1+i &&& \\
&&&&& 1-i &&&&&& 1+i &&&& \\
&&&&&& 1-i &&&& 1+i &&&&& \\
&&&&&&& 1-i && 1+i &&&&&& \\
&&&&&&&& 2 &&&&&&& \\
&&&&&&& 1+i && 1-i &&&&&& \\
&&&&&& 1+i &&&& 1-i &&&&& \\
&&&&& 1+i &&&&&& 1-i &&&& \\
&&&& 1+i &&&&&&&& 1-i &&& \\
&&& 1+i &&&&&&&&&& 1-i && \\
&& 1+i &&&&&&&&&&&& 1-i & \\
& 1+i &&&&&&&&&&&&&& 1-i
\end{bmatrix}.$$

Thus the DFT $\mathbf{F}$ is expressible in terms of the Hartley factorization by

$$\mathbf{F} = \mathbf{\Phi L_4 L_3 L_2 L_1 P_{16} f}.$$

Since $\mathbf{F} = \mathbf{Wf}$ it follows that

$$\mathbf{W} = \mathbf{\Phi L_4 L_3 L_2 L_1 P_{16}}.$$

This is a new factorization of the discrete Fourier transform matrix $\boldsymbol{W}$ and is the basis of the fast algorithm described in the next chapter.

Problems

7.1 *Factorization for FFT.* In the case of $N = 4$ show that

$$\mathrm{W} = \begin{bmatrix} 1 & 0 & 1 & 0 \\ 0 & 1 & 0 & W \\ 1 & 0 & W^2 & 0 \\ 0 & 1 & 0 & W^3 \end{bmatrix} \begin{bmatrix} 1 & 1 & 0 & 0 \\ 1 & W^2 & 0 & 0 \\ 0 & 0 & 1 & 1 \\ 0 & 0 & 1 & W^2 \end{bmatrix} \begin{bmatrix} 1 & 0 & 0 & 0 \\ 0 & 0 & 1 & 0 \\ 0 & 1 & 0 & 0 \\ 0 & 0 & 0 & 1 \end{bmatrix} \begin{bmatrix} f(0) \\ f(1) \\ f(2) \\ f(3) \end{bmatrix}.$$

7.2 *Converse of* $\mathbf{\Phi}$. Give the matrix that converts a 16-element DFT to the DHT of the same original input $\mathbf{f}$.

7.3 *Sequence of factors.* The operator $\mathbf{L}_4\mathbf{L}_3\mathbf{L}_2\mathbf{L}_1$ is applied to $\mathbf{f}$ without first permuting. Is it true that the resulting sequence consists of correct values of $\mathbf{H}$ but in an incorrect order?

7.4 *Scatter diagram.* (a) Show that the mean values $< i >$ and $< j >$ both equal $\frac{1}{2}(N-1) = m$. (b) Evaluate the moments $< (i-m)^2 >$ and $< (j-m)^2 >$. (c) Show that the correlation coefficient $< (i-m)(j-m) >$ of a permutation scatter diagram is rather low, being less than one per cent for $N > 1024$.

7.5 *Permutation on a ring.* (a) Draw N small circles equally spaced on a much bigger circle. Draw a line from the small circle to point at azimuth $j2\pi/N$, where $j = P_N(i)$, for all i. If $j \neq i$, fill in the small circle. Do this for $N = 8$, 16, 32 and 64. (b) How many axes of symmetry are there?

7.6 *Percentage of swaps.* By counting the lines in the ring diagram of the previous problem, or otherwise, show that the number of swapping operations divided by N is $0.5 - 2^{-\mathrm{IP}[(P+1)/2]}$, where $\mathrm{IP}[x]$ is the integral part of x.

CHAPTER 8

THE FAST ALGORITHM

"Through the stringybarks and saplings, on the rough and broken ground,
Down the hillside at a racing pace he went;
And he never drew the bridle till he landed safe and sound
At the bottom of that terrible descent."
A.B. ("Banjo") Paterson, "The Man from Snowy River"

We have seen that a real discrete transform can be defined that is analogous to the discrete Fourier transform; now we show that there is a fast way of computing that is analogous to the Fast Fourier Transform. As has been shown in the discussion of matrix factorization it is a simple step to go from the discrete Hartley transform to the discrete Fourier transform. Consequently the fast procedure discussed below also constitutes a fast way of arriving at the discrete Fourier transform. In fact the approach via the Hartley transform proves to be advantageous, mainly because of the simplification that results from the fact that no complex arithmetic is required when real data are being processed.

The definition equations

When the summation defining the discrete Hartley transform is expanded there are N equations and the right hand sides comprise N terms each. The four equations given earlier for the discrete Fourier transform to illustrate this feature are rewritten here in the form applying to the discrete Hartley transform. Thus

$$H(\nu) = N^{-1} \sum_{t=0}^{N=1} f(\tau) \operatorname{cas} 2\pi\nu t/N$$

expands into

$$H(0) = \tfrac{1}{4}[f(0) + f(1) + f(2) + f(3)]$$
$$H(1) = \tfrac{1}{4}[f(0) + f(1) \operatorname{cas} 2\pi/N + f(2) \operatorname{cas} 2\pi 2/N + f(3) \operatorname{cas} 2\pi 3/N]$$
$$H(2) = \tfrac{1}{4}[f(0) + f(1) \operatorname{cas} 2\pi 2/N + f(2) \operatorname{cas} 2\pi 4/N + f(3) \operatorname{cas} 2\pi 6/N]$$
$$H(3) = \tfrac{1}{4}[f(0) + f(1) \operatorname{cas} 2\pi 3/N + f(2) \operatorname{cas} 2\pi 6/N + f(3) \operatorname{cas} 2\pi 9/N].$$

For $N = 2$ there are only two equations, which reduce to

$$H(0) = \tfrac{1}{2}[f(0) + f(1)]$$

$$H(1) = \tfrac{1}{2}[f(0) - f(1)].$$

When $N = 4$ and actual numerical values are substituted for the cas functions the four equations become

$$\begin{aligned}
H(0) &= \tfrac{1}{4}[f(0) + f(1) + f(2) + f(3)] \\
H(1) &= \tfrac{1}{4}[f(0) + f(1) - f(2) - f(3)] \\
H(2) &= \tfrac{1}{4}[f(0) - f(1) + f(2) - f(3)] \\
H(3) &= \tfrac{1}{4}[f(0) - f(1) - f(2) + f(3)].
\end{aligned}$$

The coefficients can be verified by reference to a graph of the cas function (Fig. 8.1) and marking off points $\oplus$ at $\frac{1}{4}$-cycle intervals.

When $N = 8$ the equations are, as may be verified from the graph,

$$\begin{aligned}
H(0) &= \tfrac{1}{8}[f(0) + f(1) + f(2) + f(3) + f(4) + f(5) + f(6) + f(7)] \\
H(1) &= \tfrac{1}{8}[f(0) + \sqrt{2}f(1) + f(2) + 0 - f(4) - \sqrt{2}f(5) - f(6) - 0] \\
H(2) &= \tfrac{1}{8}[f(0) + f(1) - f(2) - f(3) + f(4) + f(5) - f(6) - f(7)] \\
H(3) &= \tfrac{1}{8}[f(0) + 0 - f(2) + \sqrt{2}f(3) - f(4) + 0 + f(6) - \sqrt{2}f(7)] \\
H(4) &= \tfrac{1}{8}[f(0) - f(1) + f(2) - f(3) + f(4) - f(5) + f(6) - f(7)] \\
H(5) &= \tfrac{1}{8}[f(0) - \sqrt{2}f(1) + f(2) + 0 - f(4) + \sqrt{2}f(5) - f(6) + 0] \\
H(6) &= \tfrac{1}{8}[f(0) - f(1) - f(2) + f(3) + f(4) - f(5) - f(6) + f(7)] \\
H(7) &= \tfrac{1}{8}[f(0) + 0 - f(2) - \sqrt{2}f(3) - f(4) + 0 + f(6) + \sqrt{2}f(7)].
\end{aligned}$$

One can count the number of multiplications and additions and use that count to determine the time that would be needed for computing all N transform values.

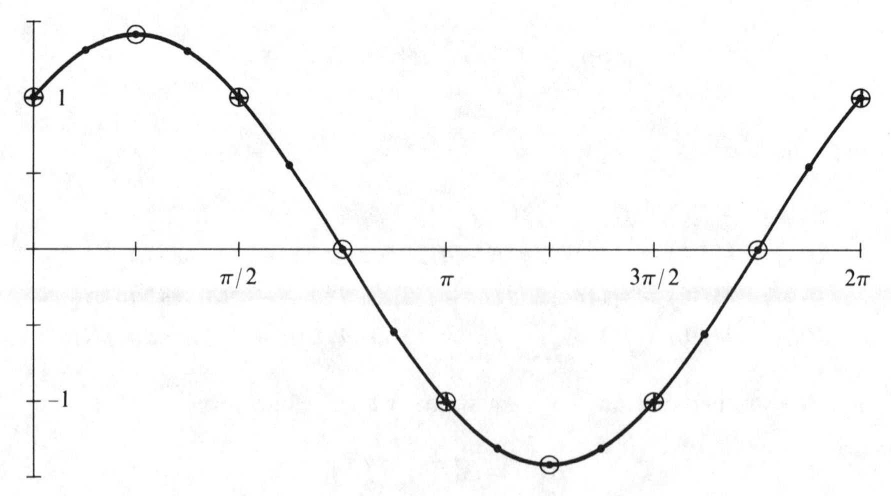

Fig. 8.1. Graph of $y = \operatorname{cas} x$ from which the coefficients can be read. For $N = 4$ use $\oplus$. For $N = 8$ include $\odot$. Points for $N = 16$ are indicated faintly.

The virtues of $N \log N$

Let the number N of elements in a sequence be 2 raised to some power P. Then $P = \log_2 N$. Logarithms to base 2 are not easily pictured but it is desirable to have a feeling for the magnitude of $\log_2 N$, which is simply the same as the power P in

$$N = 2^P.$$

Table 8.1 shows the association of N with P over the range of practical interest.

It is useful to memorize some of the entries in this table, particularly the rows emphasized; adjacent entries will then be on the tip of the tongue. If you do not know that 2^{10} is about 1000 you will be left behind in some technical conversations. Another useful milestone: 2^{32} bits $= 4 \times 10^9$ bits $= 500$ megabytes. The values of N are shown with factors, in some cases, as a reminder of N-values that correspond to certain image sizes. At $P = 18$, N is about a quarter of a million and about equal to the number of pixels in a television image.

Table 8.1 Values of P, N and NP

P	N	NP
1	2	2
2	4	8
3	8	24
4	16	64
5	32	160
6	**64 = 8 × 8**	**384**
7	128	896
8	256 = 16 × 16	2048
9	512	4608
10	**1024 = 32 × 32**	**10,240**
11	2048	22,528
12	4096 = 64 × 64	49,152
13	8192	106,496
14	16,384	229,376
15	32,768	491,520
16	**65,536**	**1,048,576**
17	131,072	2,228,224
18	262,144 = 512 × 512	4,718,592
19	524,288	9,961,472
20	**1,048,576 = 1024 × 1024**	**20,971,520**

It is commonly said that the number of operations in the equations above increases as N^2. This is manifestly not the case for $N = 2$, 4 and 8. For large values of N a complication in counting arises from the fact that a noticeable number of coefficients assume a value of unity and do not require a multiplication; other coefficients are zero, which eliminates both a multiplication and a division. Despite the complication, the coefficients are determinate and can clearly be counted for successively greater values of N. Then the computing time can be arrived at by allowing appropriate times per addition and per multiplication and it is generally thought that the total time is proportional to N^2. Of course there is also the possibility of timing actual computations, an approach which goes beyond counting and can encompass time devoted to necessary steps that are not included in the sort of operations count described above.

As N becomes large, dependence on N^2 means that practical limits of time or cost will be encountered. A fast algorithm is one that does not evaluate the summations specified by the definition but instead requires times proportional to NP, or $N\log_2 N$, instead of N^2. As an illustration of the practical significance consider a fast algorithm that consumes $10NP$ microseconds compared with $5N^2$ microseconds. Fig. 8.2 covers a range

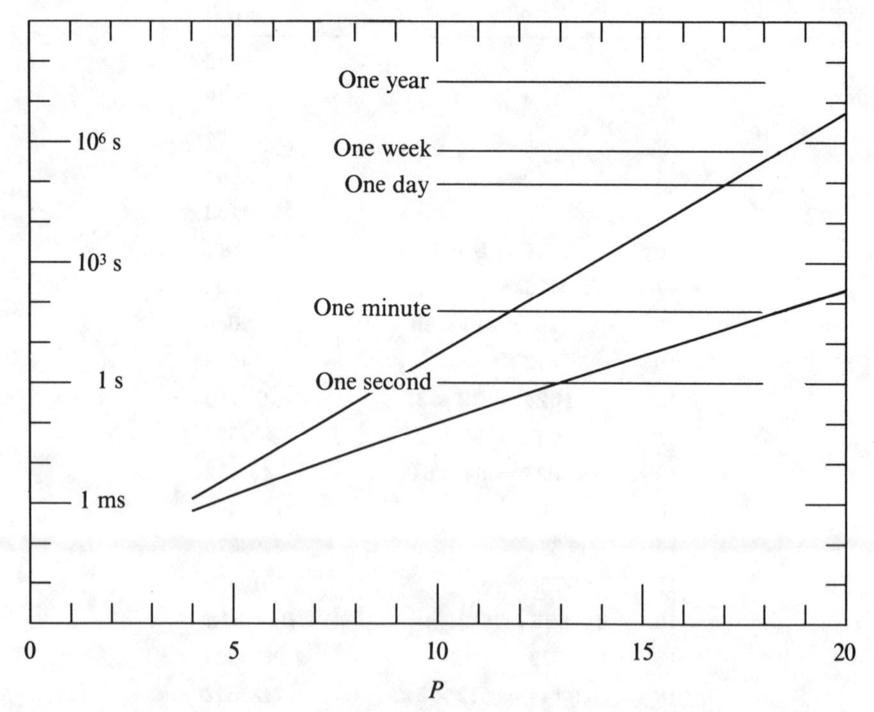

Fig. 8.2. Logarithmic plots of $10NP$ μs (below) and $5N^2$ μs (above) versus $P = \log_2 N$ showing how fast algorithms have opened up the range $10 < P < 20$ for general applications.

up to $P = 20$ $(N = 1{,}048{,}576)$. While this is not the largest practical value of P it is what would be met in an image of 1024×1024 pixels, which is somewhat higher resolution than a television screen offers. The great majority of transforms that are performed each day are undoubtedly in the range up to $P = 10$. The graph shows dramatically that, as larger images are processed digitally, the speed of the algorithm becomes vital. In fact it is clear that the introduction of the Fast Fourier Transform (FFT) opened up possibilities that could not have been contemplated without it. The need for image processing in real time calls for even faster speeds today and conversely, as better speeds are attained, new applications will be made possible.

Repeated halving

If the Fourier transform of a given sequence is needed it is possible to split the sequence into two parts and transform the parts separately. Because the time taken increases faster than N there will be a time saving as regards transformation; but some additional time will be used up as the two short transforms are somehow combined to assemble the longer transform. The idea sounds attractive because it can be applied again – the short sequences themselves can be halved for a further time saving. Indeed, any factorization at all will diminish the computing load. When the sequence has been halved $P - 1$ times, the length of the remaining segments is two elements. Now transforming a two-element sequence is trivial, involving only two additions and no multiplications. The number of operations required to combine the outputs at each stage is of the order of N because some coefficient has to be attached to each of N elements, and there are $P - 1$ stages. Consequently as N gets larger the number of operations goes as $N(P - 1)$, or NP approximately.

As with the justification of the N^2 rule, the argument leading to NP is only rough. We now look in detail at the mechanism for building up the Hartley transform. Later we look into the running time empirically in an investigation that supersedes the operation counting procedure.

Transformation by decomposition

Consider a data sequence

$$\begin{aligned} f(\tau) &= \{a_1 \quad a_2 \quad b_1 \quad b_2 \quad c_1 \quad c_2 \quad d_1 \quad d_2\} \\ &= \{a_1 \quad 0 \quad b_1 \quad 0 \quad c_1 \quad 0 \quad d_1 \quad 0\} + \{0 \quad a_2 \quad 0 \quad b_2 \quad 0 \quad c_2 \quad 0 \quad d_2\} \end{aligned}$$

which is shown as the sum of two parts. The two short sequences $\{a_1 \quad b_1 \quad c_1 \quad d_1\}$ and $\{a_2 \quad b_2 \quad c_2 \quad d_2\}$, if interleaved precisely, as in the perfect shuffle, would reconstitute the original sequence $f(\tau)$.

We wish to find the Hartley transform $H(\nu)$ of $f(\tau)$, and to do this from the transforms of the short sequences. Suppose that

$$\{a_1 \quad b_1 \quad c_1 \quad d_1\} \quad \text{has DHT} \quad \{\alpha_1 \quad \beta_1 \quad \gamma_1 \quad \delta_1\}$$

and that

$$\{a_2 \quad b_2 \quad c_2 \quad d_2\} \text{ has DHT } \{\alpha_2 \quad \beta_2 \quad \gamma_2 \quad \delta_2\}.$$

From the stretch theorem for the DHT we can say immediately that

$$\{a_1\ 0\ b_1\ 0\ c_1\ 0\ d_1\ 0\} \quad \text{has DHT} \quad \{\alpha_1\ \beta_1\ \gamma_1\ \delta_1\ \alpha_1\ \beta_1\ \gamma_1\ \delta_1\}$$

and

$$\{a_2\ 0\ b_2\ 0\ c_2\ 0\ d_2\ 0\} \quad \text{has DHT} \quad \{\alpha_2\ \beta_2\ \gamma_2\ \delta_2\ \alpha_2\ \beta_2\ \gamma_2\ \delta_2\}.$$

However, we require the DHT of $\{0 \quad a_2 \quad 0 \quad b_2 \quad 0 \quad c_2 \quad 0 \quad d_2\}$. To get this we apply the shift theorem. When the shift is one element to the right the shift theorem reads as follows.

If $g(\tau)$ has DHT $G(\nu)$ then $g(\tau - 1)$ has DHT $G(\nu)\cos(2\pi\nu/N) + G(N-\nu)\sin(2\pi\nu/N)$.

Writing $\Theta = 2\pi/N$, we can say that the sequence $\{G(\nu)\}$ becomes $\{G(0) \quad G(1)\cos\Theta \quad G(2)\cos 2\Theta \quad G(3)\cos 3\Theta \quad G(4)\cos 4\Theta \quad G(5)\cos 5\Theta \quad G(6)\cos 6\Theta \quad G(7)\sin 7\Theta\} \quad + \quad \{G(7)\sin\Theta \quad G(6)\sin 2\Theta \quad G(5)\sin 3\Theta \quad G(4)\sin 4\Theta \quad G(3)\sin 5\Theta \quad G(2)\sin 6\Theta \quad G(1)\sin 7\Theta \quad 0\}$.

Consequently, given that

$$g(\tau) = \{a_2 \quad 0 \quad b_2 \quad 0 \quad c_2 \quad 0 \quad d_2 \quad 0\}$$

has a DHT

$$G(\nu) = \{\alpha_2 \quad \beta_2 \quad \gamma_2 \quad \delta_2 \quad \alpha_2 \quad \beta_2 \quad \gamma_2 \quad \delta_2\}$$

then

$$g(\tau - 1) = \{0 \quad a_2 \quad 0 \quad b_2 \quad 0 \quad c_2 \quad 0 \quad d_2\} \quad \text{has DHT}$$

$$\{\alpha_2 \quad \beta_2\cos\Theta \quad \gamma_2\cos 2\Theta \quad \delta_2\cos 3\Theta \quad \alpha_2\cos 4\Theta \quad \beta_2\cos 5\Theta \quad \gamma_2\cos 6\Theta\} \; \delta_2\cos 7\Theta\} + \{0 \quad \delta_2\sin\Theta \quad \gamma_2\sin 2\Theta \quad 0 \quad \delta_2\sin 5\Theta \quad \gamma_2\sin 6\Theta \quad \beta_2\sin 7\Theta\}.$$

Finally, applying the addition theorem, we have

$$H(\nu) = \{\alpha_1 \quad \beta_1 \quad \gamma_1 \quad \delta_1 \quad \alpha_1 \quad \beta_1 \quad \gamma_1 \quad \delta_1\}$$
$$+\{\alpha_2 \quad \beta_2\cos\Theta \quad \gamma_2\cos 2\Theta \quad \delta_2\cos 3\Theta \quad \alpha_2\cos 4\Theta \quad \beta_2\cos 5\Theta \quad \gamma_2\cos 6\Theta$$
$$\delta_2\cos 7\Theta\} + \{0 \quad \delta_2\sin\Theta \quad \gamma_2\sin 2\Theta \quad 0 \quad \delta_2\sin 5\Theta \quad \gamma_2\sin 6\Theta \quad \beta_2\sin 7\Theta\}.$$

If in this equation for $N = 8$ the factors that are 0, 1 or -1 are entered as such the result simplifies to

$$H(\nu) = \{\alpha_1 + \alpha_2 \quad \beta_1 + r(\beta_2 + \delta_2) \quad \gamma_1 + \gamma_2 \quad \alpha_1 + \alpha_2$$
$$\beta_1 - r(\beta_2 + \delta_2) \quad \gamma_1 - \gamma_2 \quad \delta_1 - r(\beta_2 - \delta_2)\}.$$

This completes the demonstration that the 8-element transform $H(\nu)$ can be built up from the two 4-element transforms $\{\alpha_1 \quad \beta_1 \quad \gamma_1 \quad \delta_1\}$ and $\{\alpha_2 \quad \beta_2 \quad \gamma_2 \quad \delta_2\}$.

The general decomposition formula

It is clear how to generalize the structure of the foregoing equation. Given $f(\tau)$, $\tau = 0, 1, \ldots N-1$, let the N-element subsequences

$$\{f(0) \quad 0 \quad f(2) \quad 0 \quad f(4) \quad 0 \quad \ldots\}, \quad \{f(1) \quad 0 \quad f(3) \quad 0 \quad f(5) \quad 0 \quad \ldots\}$$

have DHTs $H_1(\nu)$ and $H_2(\nu)$ respectively. Then

$$H(\nu) = H_1(\nu) + H_2(\nu)\cos(2\pi\nu/N) + H_2(N-\nu)\sin(2\pi\nu/N).$$

This is the general decomposition formula that produces the discrete Hartley transform by successive halving.

Since $H_1(\nu)$ and $H_2(\nu)$ can themselves be derived by continued decomposition until four-element sequences are reached, the complete breakdown is expressible in terms of the original data. For example, in a simple 8-element transform, the element $H(3)$ is found to be related to the eight elements of $f(\tau)$ as follows:

$$H(3) = \tfrac{1}{8}[f(0) + 0 - f(2) + \sqrt{2}f(3) - f(4) + 0 + f(6) - \sqrt{2}f(7)].$$

This dependence is consistent with

$$H(3) = \tfrac{1}{8}\{f(0) + f(1)\,\mathrm{cas}\,3\Theta + f(2)\,\mathrm{cas}\,6\Theta + f(3)\,\mathrm{cas}\,9\Theta + f(4)\,\mathrm{cas}\,12\Theta + f(5)\,\mathrm{cas}\,15\Theta + f(6)\,\mathrm{cas}\,18\Theta + f(7)\,\mathrm{cas}\,21\Theta\},$$

which comes directly from the transform definition. It is interesting to ask why the two approaches should take different times to compute. The difference lies in the organization into levels that the successive decompositions permit. This structure, although arrived at independently in the present development, corresponds to the factorization of the matrix.

To see the sequence of the arithmetic operations, one turns to Table 8.2 where the actual equations for implementing the transformation are shown.

Equations for $N = 16$

As with the matrix factorization, all the features can be brought out by the example of $N = 16$. In Table 8.2 the first column gives the original data $F(0,\nu)$. The second column contains the same data permuted as defined in an earlier chapter. In the subsequent columns $F(s,\nu)$ stands for the νth element of a 16-element array at stage s of its transformation. Thus the column $F(1,\nu)$ is the first stage. Arrows have been used in the assignment statements to emphasize the direction of flow. We see that in all cases $F(1,\nu)$ is obtained by a simple sum or difference of two data elements. Stage 2 then operates on these combinations by adding together three terms

Table 8.2 Equations relating successive stages of the 16-element FHT as indexed by stage number s

Data	Permute	Stage 1	Stage 2
$F(0,0)$	$F(0,0) \rightarrow F(0,0)$	$F(0,0) + F(0,1) \rightarrow F(1,0)$	$F(1,0) + F(1,2)C_0 + F(1,2)S_0 \rightarrow F(2,0)$
$F(0,1)$	$F(0,1) \rightarrow F(0,8)$	$F(0,0) - F(0,1) \rightarrow F(1,1)$	$F(1,1) + F(1,3)C_1 + F(1,3)S_1 \rightarrow F(2,1)$
$F(0,2)$	$F(0,2) \rightarrow F(0,4)$	$F(0,2) + F(0,3) \rightarrow F(1,2)$	$F(1,0) + F(1,2)C_2 + F(1,2)S_2 \rightarrow F(2,2)$
$F(0,3)$	$F(0,3) \rightarrow F(0,12)$	$F(0,2) - F(0,3) \rightarrow F(1,3)$	$F(1,1) + F(1,3)C_3 + F(1,3)S_3 \rightarrow F(2,3)$
$F(0,4)$	$F(0,4) \rightarrow F(0,2)$	$F(0,4) + F(0,5) \rightarrow F(1,4)$	$F(1,4) + F(1,6)C_0 + F(1,6)S_0 \rightarrow F(2,4)$
$F(0,5)$	$F(0,5) \rightarrow F(0,10)$	$F(0,4) - F(0,5) \rightarrow F(1,5)$	$F(1,5) + F(1,7)C_1 + F(1,7)S_1 \rightarrow F(2,5)$
$F(0,6)$	$F(0,6) \rightarrow F(0,6)$	$F(0,6) + F(0,7) \rightarrow F(1,6)$	$F(1,4) + F(1,6)C_2 + F(1,6)S_2 \rightarrow F(2,6)$
$F(0,7)$	$F(0,7) \rightarrow F(0,14)$	$F(0,6) - F(0,7) \rightarrow F(1,7)$	$F(1,5) + F(1,7)C_3 + F(1,7)S_3 \rightarrow F(2,7)$
$F(0,8)$	$F(0,8) \rightarrow F(0,1)$	$F(0,8) + F(0,9) \rightarrow F(1,8)$	$F(1,8) + F(1,10)C_0 + F(1,10)S_0 \rightarrow F(2,8)$
$F(0,9)$	$F(0,9) \rightarrow F(0,9)$	$F(0,8) - F(0,9) \rightarrow F(1,9)$	$F(1,9) + F(1,11)C_1 + F(1,11)S_1 \rightarrow F(2,9)$
$F(0,10)$	$F(0,10) \rightarrow F(0,5)$	$F(0,10) + F(0,11) \rightarrow F(1,10)$	$F(1,8) + F(1,10)C_2 + F(1,10)S_2 \rightarrow F(2,10)$
$F(0,11)$	$F(0,11) \rightarrow F(0,13)$	$F(0,10) - F(0,11) \rightarrow F(1,11)$	$F(1,9) + F(1,11)C_3 + F(1,11)S_3 \rightarrow F(2,11)$
$F(0,12)$	$F(0,12) \rightarrow F(0,3)$	$F(0,12) + F(0,13) \rightarrow F(1,12)$	$F(1,12) + F(1,14)C_0 + F(1,14)S_0 \rightarrow F(2,12)$
$F(0,13)$	$F(0,13) \rightarrow F(0,11)$	$F(0,12) - F(0,13) \rightarrow F(1,13)$	$F(1,13) + F(1,15)C_1 + F(1,15)S_1 \rightarrow F(2,13)$
$F(0,14)$	$F(0,14) \rightarrow F(0,7)$	$F(0,14) + F(0,15) \rightarrow F(1,14)$	$F(1,12) + F(1,14)C_2 + F(1,14)S_2 \rightarrow F(2,14)$
$F(0,15)$	$F(0,15) \rightarrow F(0,15)$	$F(0,14) - F(0,15) \rightarrow F(1,15)$	$F(1,13) + F(1,15)C_3 + F(1,15)S_3 \rightarrow F(2,15)$

Stage 3	16× DHT
$\boldsymbol{F}(2,0) + \boldsymbol{F}(2,4)C_0 + \boldsymbol{F}(2,4)S_0 \rightarrow \boldsymbol{F}(3,0)$	$\boldsymbol{F}(3,0) + \boldsymbol{F}(3,8)C_0 + \boldsymbol{F}(3,8)S_0 \rightarrow \boldsymbol{F}(4,0)$
$\boldsymbol{F}(2,1) + \boldsymbol{F}(2,5)C_1 + \boldsymbol{F}(2,7)S_1 \rightarrow \boldsymbol{F}(3,1)$	$\boldsymbol{F}(3,1) + \boldsymbol{F}(3,9)C_1 + \boldsymbol{F}(3,15)S_1 \rightarrow \boldsymbol{F}(4,1)$
$\boldsymbol{F}(2,2) + \boldsymbol{F}(2,6)C_2 + \boldsymbol{F}(2,6)S_2 \rightarrow \boldsymbol{F}(3,2)$	$\boldsymbol{F}(3,2) + \boldsymbol{F}(3,10)C_2 + \boldsymbol{F}(3,14)S_2 \rightarrow \boldsymbol{F}(4,2)$
$\boldsymbol{F}(2,3) + \boldsymbol{F}(2,7)C_3 + \boldsymbol{F}(2,5)S_3 \rightarrow \boldsymbol{F}(3,3)$	$\boldsymbol{F}(3,3) + \boldsymbol{F}(3,11)C_3 + \boldsymbol{F}(3,13)S_3 \rightarrow \boldsymbol{F}(4,3)$
$\boldsymbol{F}(2,0) + \boldsymbol{F}(2,4)C_4 + \boldsymbol{F}(2,4)S_4 \rightarrow \boldsymbol{F}(3,4)$	$\boldsymbol{F}(3,4) + \boldsymbol{F}(3,12)C_4 + \boldsymbol{F}(3,12)S_4 \rightarrow \boldsymbol{F}(4,4)$
$\boldsymbol{F}(2,1) + \boldsymbol{F}(2,5)C_5 + \boldsymbol{F}(2,7)S_5 \rightarrow \boldsymbol{F}(3,5)$	$\boldsymbol{F}(3,5) + \boldsymbol{F}(3,13)C_5 + \boldsymbol{F}(3,11)S_5 \rightarrow \boldsymbol{F}(4,5)$
$\boldsymbol{F}(2,2) + \boldsymbol{F}(2,6)C_6 + \boldsymbol{F}(2,6)S_6 \rightarrow \boldsymbol{F}(3,6)$	$\boldsymbol{F}(3,6) + \boldsymbol{F}(3,14)C_6 + \boldsymbol{F}(3,10)S_6 \rightarrow \boldsymbol{F}(4,6)$
$\boldsymbol{F}(2,3) + \boldsymbol{F}(2,7)C_7 + \boldsymbol{F}(2,5)S_7 \rightarrow \boldsymbol{F}(3,7)$	$\boldsymbol{F}(3,7) + \boldsymbol{F}(3,15)C_7 + \boldsymbol{F}(3,9)S_7 \rightarrow \boldsymbol{F}(4,7)$
$\boldsymbol{F}(2,8) + \boldsymbol{F}(2,12)C_0 + \boldsymbol{F}(2,12)S_0 \rightarrow \boldsymbol{F}(3,8)$	$\boldsymbol{F}(3,0) + \boldsymbol{F}(3,8)C_8 + \boldsymbol{F}(3,8)S_8 \rightarrow \boldsymbol{F}(4,8)$
$\boldsymbol{F}(2,9) + \boldsymbol{F}(2,13)C_1 + \boldsymbol{F}(2,15)S_1 \rightarrow \boldsymbol{F}(3,9)$	$\boldsymbol{F}(3,1) + \boldsymbol{F}(3,9)C_9 + \boldsymbol{F}(3,15)S_9 \rightarrow \boldsymbol{F}(4,9)$
$\boldsymbol{F}(2,10) + \boldsymbol{F}(2,14)C_2 + \boldsymbol{F}(2,14)S_2 \rightarrow \boldsymbol{F}(3,10)$	$\boldsymbol{F}(3,2) + \boldsymbol{F}(3,10)C_{10} + \boldsymbol{F}(3,14)S_{10} \rightarrow \boldsymbol{F}(4,10)$
$\boldsymbol{F}(2,11) + \boldsymbol{F}(2,15)C_3 + \boldsymbol{F}(2,13)S_3 \rightarrow \boldsymbol{F}(3,11)$	$\boldsymbol{F}(3,3) + \boldsymbol{F}(3,11)C_{11} + \boldsymbol{F}(3,13)S_{11} \rightarrow \boldsymbol{F}(4,11)$
$\boldsymbol{F}(2,8) + \boldsymbol{F}(2,12)C_4 + \boldsymbol{F}(2,12)S_4 \rightarrow \boldsymbol{F}(3,12)$	$\boldsymbol{F}(3,4) + \boldsymbol{F}(3,12)C_{12} + \boldsymbol{F}(3,12)S_{12} \rightarrow \boldsymbol{F}(4,12)$
$\boldsymbol{F}(2,9) + \boldsymbol{F}(2,13)C_5 + \boldsymbol{F}(2,15)S_5 \rightarrow \boldsymbol{F}(3,13)$	$\boldsymbol{F}(3,5) + \boldsymbol{F}(3,13)C_{13} + \boldsymbol{F}(3,11)S_{13} \rightarrow \boldsymbol{F}(4,13)$
$\boldsymbol{F}(2,10) + \boldsymbol{F}(2,14)C_6 + \boldsymbol{F}(2,14)S_6 \rightarrow \boldsymbol{F}(3,14)$	$\boldsymbol{F}(3,6) + \boldsymbol{F}(3,14)C_{14} + \boldsymbol{F}(3,10)S_{14} \rightarrow \boldsymbol{F}(4,14)$
$\boldsymbol{F}(2,11) + \boldsymbol{F}(2,15)C_7 + \boldsymbol{F}(2,13)S_7 \rightarrow \boldsymbol{F}(3,15)$	$\boldsymbol{F}(3,7) + \boldsymbol{F}(3,15)C_{15} + \boldsymbol{F}(3,9)S_{15} \rightarrow \boldsymbol{F}(4,15)$

taken from the decomposition formula and stages 3 and 4 follow suit. The retrograde indexing is clearly seen in the values of $F(3, \nu)$ associated with sine factors.

The number of multiplications is now seen to be economical by comparison with the N^2 multiplications required to evaluate the definition summation. However, this table does not lend itself to an accurate count because of the degeneracy, well illustrated in Table 8.3, that is evident when numerical values are substituted for the trigonometric coefficients. In neither stage 1 nor stage 2 is any multiplication required at all and in stage 4 only 16 multiplications are needed, all of which use the same factor $r = 1/\sqrt{2}$. The integrated effect of the deterministic but involved simplification is given later in timing studies.

Flow diagram

One could trace lines through the tables showing the way the influence of each input element of data is felt in each output transform value. Such a flow diagram for $N = 8$ is shown in Fig. 8.3. First there is a box labeled PERMUTE whose contents will be examined later; it simply rearranges the order of data. The stage 1 boxes take two inputs and give two outputs; they perform the DHT on a sequence with $N = 2$, the operation described by

$$H(\nu) = \tfrac{1}{2}\sum_{\tau=0}^{1} f(\tau)\operatorname{cas} 2\pi\nu/2.$$

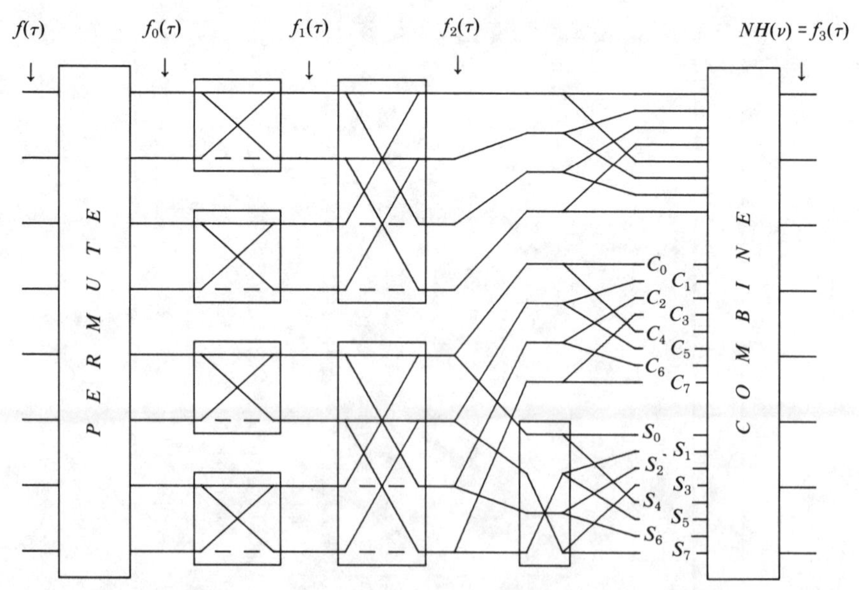

Fig. 8.3. Each input to the fast algorithm contributes to each output via flow lines associated with transfer factors C_n and S_n which in special cases reduce to 1 (solid line) or -1 (dashed line).

Table 8.3 The 16-element equations with numerical coefficients inserted. The factor r is $1/\sqrt{2}$

Data	Permute	Stage 1
$F(0,0)$	$F(0,0) \to F(0,0)$	$F(0,0) + F(0,1) \to F(1,0)$
$F(0,1)$	$F(0,8) \to F(0,1)$	$F(0,0) - F(0,1) \to F(1,1)$
$F(0,2)$	$F(0,4) \to F(0,2)$	$F(0,2) + F(0,3) \to F(1,2)$
$F(0,3)$	$F(0,12) \to F(0,3)$	$F(0,2) - F(0,3) \to F(1,3)$
$F(0,4)$	$F(0,2) \to F(0,4)$	$F(0,4) + F(0,5) \to F(1,4)$
$F(0,5)$	$F(0,10) \to F(0,5)$	$F(0,4) - F(0,5) \to F(1,5)$
$F(0,6)$	$F(0,6) \to F(0,6)$	$F(0,6) + F(0,7) \to F(1,6)$
$F(0,7)$	$F(0,14) \to F(0,7)$	$F(0,6) - F(0,7) \to F(1,7)$
$F(0,8)$	$F(0,1) \to F(0,8)$	$F(0,8) + F(0,9) \to F(1,8)$
$F(0,9)$	$F(0,9) \to F(0,9)$	$F(0,8) - F(0,9) \to F(1,9)$
$F(0,10)$	$F(0,5) \to F(0,10)$	$F(0,10) + F(0,11) \to F(1,10)$
$F(0,11)$	$F(0,13) \to F(0,11)$	$F(0,10) - F(0,11) \to F(1,11)$
$F(0,12)$	$F(0,3) \to F(0,12)$	$F(0,12) + F(0,13) \to F(1,12)$
$F(0,13)$	$F(0,11) \to F(0,13)$	$F(0,12) - F(0,13) \to F(1,13)$
$F(0,14)$	$F(0,7) \to F(0,14)$	$F(0,14) + F(0,15) \to F(1,14)$
$F(0,15)$	$F(0,15) \to F(0,15)$	$F(0,14) - F(0,15) \to F(1,15)$

Stage 2	Stage 3
$F(1,0) + F(1,2) \to F(2,0)$	$F(2,0) \qquad + F(2,4) \to F(3,0)$
$F(1,1) + F(1,3) \to F(2,1)$	$F(2,1) + rF(2,5) + rF(2,7) \to F(3,1)$
$F(1,0) - F(1,2) \to F(2,2)$	$F(2,2) \qquad + F(2,6) \to F(3,2)$
$F(1,1) - F(1,3) \to F(2,3)$	$F(2,3) - rF(2,7) + rF(2,5) \to F(3,3)$
$F(1,4) + F(1,6) \to F(2,4)$	$F(2,0) \qquad - F(2,4) \to F(3,4)$
$F(1,5) + F(1,7) \to F(2,5)$	$F(2,1) - rF(2,5) - rF(2,7) \to F(3,5)$
$F(1,4) - F(1,6) \to F(2,6)$	$F(2,2) \qquad - F(2,6) \to F(3,6)$
$F(1,5) - F(1,7) \to F(2,7)$	$F(2,3) + rF(2,7) - rF(2,5) \to F(3,7)$
$F(1,8) + F(1,10) \to F(2,8)$	$F(2,8) \qquad + F(2,12) \to F(3,8)$
$F(1,9) + F(1,11) \to F(2,9)$	$F(2,9) + rF(2,13) + rF(2,15) \to F(3,9)$
$F(1,8) - F(1,10) \to F(2,10)$	$F(2,10) \qquad + F(2,14) \to F(3,10)$
$F(1,9) - F(1,11) \to F(2,11)$	$F(2,11) - rF(2,15) + rF(2,13) \to F(3,11)$
$F(1,12) + F(1,14) \to F(2,12)$	$F(2,8) \qquad - F(2,12) \to F(3,12)$
$F(1,13) + F(1,15) \to F(2,13)$	$F(2,9) - rF(2,13) + rF(2,15) \to F(3,13)$
$F(1,12) - F(1,14) \to F(2,14)$	$F(2,10) \qquad - F(2,14) \to F(3,14)$
$F(1,13) - F(1,15) \to F(2,15)$	$F(2,11) + rF(2,15) - rF(2,13) \to F(3,15)$

Expanding this special case of the definition formula we have

$$H(0) = \tfrac{1}{2}[f(0) + f(1)]$$

$$H(1) = \tfrac{1}{2}[f(0) - f(1)].$$

The factors $\frac{1}{2}$ are not included in the flow diagram; a further factor $\frac{1}{2}$ is omitted in both subsequent stages with the consequence that a factor 8 accumulates. In practice, this factor is included with any other scale factor that is needed for normalizing, correcting or plotting, in order to avoid numerous multiplications in the intermediate stages.

As a result of the carrying over of the factors $\frac{1}{2}$ to the end, the box for stage 1 merely conducts its two inputs to a summing junction at its top output and to a differencing junction at its other output. The diagram can dispense with differencing junctions if dashed lines are introduced to signify a transmission factor -1. The stage 2 boxes take four inputs, as revealed by the tabulated equations which are the basis for the flow diagram, and produce their four outputs without need of multiplication, as already established. There are two dashed lines and six full lines. Only in stage 3 do we see the retrograde indexing beginning to appear.

The four element transform emerging from the upper stage-2 box is replicated to form the 8-element sequence previously referred to as

$$H_1(\nu) = \{\alpha_1 \quad \beta_1 \quad \gamma_1 \quad \delta_1 \quad \alpha_1 \quad \beta_1 \quad \gamma_1 \quad \delta_1\}$$

and is connected to output junctions 0 to 7. The four-element transform emerging from the lower stage-2 box is divided into two channels, one for cosines and one for sines. For the cosine channel the four values are replicated, as for the upper box, before cosine coefficients C_0 to C_7 are applied; then the products are connected to summing junctions at outputs 0 to 7. For the sine channel a cross-over box is required before replication in order to accommodate the required retrograde indexing. After application of the sine coefficients, the products are connected to the same eight output summing junctions contained in the manifold labeled COMBINE.

Computing in place

In describing the flow of operations it was tacitly understood that an N-element array was ready to receive the output from each stage as it was computed. This approach simplifies the explanation and is implemented in the demonstration programs **FHTBAS** and **FHTBAS.FOR** of Appendix 1. However, it is possible to be less wasteful of computer memory. It may have been noticed that in the tables the array $F(0,\nu)$ that contained the input data was also used to contain the recorded data after permutation; thus memory space was saved. Some juggling is required to do this. For example, if $F(0,8)$ is to be placed in location (0,1) as called for in the second row of the permute column, the value F(0,1) must first be placed in a temporary location. Later it can be placed in location (0,8) as called for in row 9.

Similar juggling is called for if the outputs of stage 1 are to be placed back in the original array. Two temporary locations will be required for each input to a stage-1 box. Four temporary locations will be required for the stage-2 boxes and so on. In the demonstration program the number of array storage locations used is NP, but in the most efficient plan the number is simply N. Then when the computation is completed the transform will occupy the original data locations. That means the original data will be obliterated; of course the original data can always be recovered by taking a further transform, but in general any operations that need to be carried out on the original data, such as tabulating or plotting, must be attended to in advance if the storage is to be restricted to just N locations.

In-place computation is utilized in **FHTSUB** and **FHTFOR.FOR** (Appendix 1). Much longer data sequences can be handled as a result of the storage locations that are freed.

Stripe diagram analysis of timing

Having found a way of getting the DHT we can inquire into the time taken to compute it as compared with the time taken for the DFT. A customary way of handling such questions is by counting operations but an alternative method is to time the computation. Actual timing analysis reveals that the different parts of the program contribute different proportions of the total running time as N changes; as a result there is no simple ratio between the speeds of the FHT and FFT programs. One may certainly expect the running time ratio to depend on N but in addition it will depend on other factors. To understand this it is necessary to see the timing analysis in a particular case.

An actual program that will be examined later in more detail has been analyzed with the results shown in the stripe diagram of Fig. 8.4. The program breaks down into a number of separate subroutines as follows: (a) overhead, (b) data insertion, (c) precomputing powers of 2, (d) precomputing sines and cosincs, (c) pcrmutation, (f) stages 1 and 2 combined, (g) stages 3 to $P-1$, (h) conversion to DFT.

Overhead refers to parts of the program that are essential but depend on N very little or perhaps not at all, and data insertion is self-explanatory. Precomputing a table of powers of two is helpful because many details of the program are concerned with component blocks of length $N/2$, $N/4$ and so on. All these numbers are powers of 2 if N is a power of 2 and it is much better to have these powers available from the beginning than to compute $N/2$ as needed (in cases where division by 2 takes a time comparable with other operations); the submultiples of N occur inside iterative loops and it is a waste of time to repeat the same division over and over again. In some computers, division by 2 may be effected in a shift register in almost zero time; in such circumstances perhaps precomputing might be reconsidered. This example illustrates the comment regarding speed ratio which is now

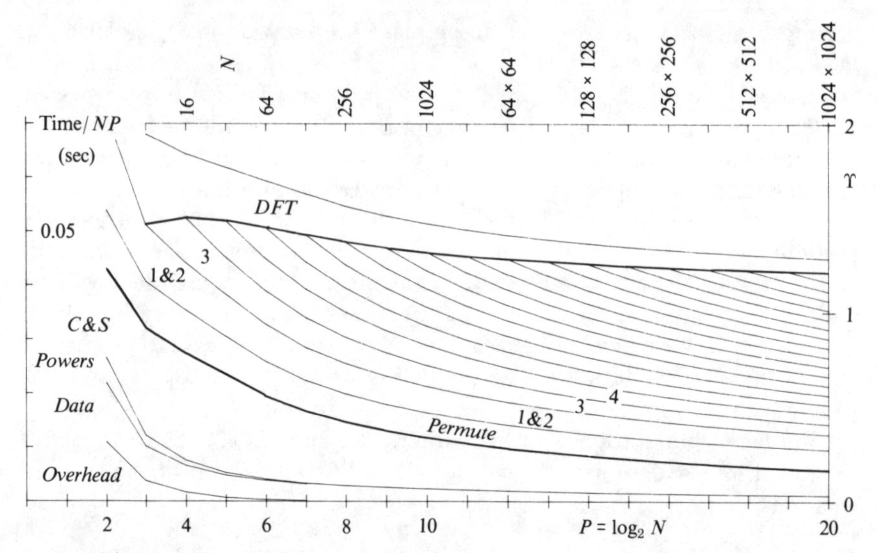

Fig. 8.4. Stripe diagram analysis of the time required to perform a DHT as a function of the length of the data sequence.

seen to be computer dependent as well as N dependent. Even if the ratio were taken on the one computer at fixed N, between two algorithms that share the same method of handling powers of 2, the speed ratio would change when the method was changed if this part of the program did not take the same fraction of time in each case.

Precomputing sines and cosines is done for the same reason, to avoid repetition, but occupies an order of magnitude more time than the powers of 2 and has attracted corresponding attention. Details of procedure will be given later.

Permutation can be seen from the diagram to require less time than precomputing trigonometric functions but the disparity diminishes with increasing length of the data sequence. It does not follow that these proportions will be the same in other cases, but the absolute values, including the dependence on N, can be ascertained by timing.

Stages 1 and 2 in the program under study were computed by the equations of Table 8.3. The alternative would be to use a general formula containing sines and cosines which, when evaluated for these two stages, would prove to have trivial values. By direct measurement it was found that the time taken for stages 1 and 2 was approximately halved if these stages were processed directly before entering the iterative loop.

The general formula was used for all subsequent stages, 3 through $P-1$, even though a small saving was found when stage 3 was handled separately. The program presented in Appendix 1 in fact repeatedly recomputes the constant $r = 0.707$ that occurs in stage 3.

Finally, a noticeable extra time is required to form the DFT, should it be needed. However, for a power spectrum there is no extra required because the power spectrum can be computed from the DHT by evaluating $[H(\nu)]^2 + [H(N-\nu)]^2$. This can be done in the same time that it takes to compute the power spectrum from the real and imaginary parts furnished by the DFT.

Naturally the time taken for the full computation increases as N increases but the various stages depend on N differently. As is well known, the running time goes roughly as $N \log N$, or NP, so for that reason the ordinate has been taken as time divided by NP. It may be seen that, as N increases

$$\text{running time} \rightarrow 0.045NP \text{ seconds.}$$

Figure of Merit

For computers with faster clock rates the constant of proportionality would be less than 0.045 but a figure of merit Υ can be defined which allows for differences between computers in the common case where multiplication time substantially exceeds the time taken for frequent operations such as addition and assignment statements. The idea is to normalize to the time T_4 taken for four multiplications. Then

$$\text{running time} = \Upsilon NPT_4.$$

The figure of merit Υ is likely to be in the neighborhood of unity and for the case before us approaches 1.3 as $P \rightarrow 20$. For shorter data sequences Υ will be somewhat larger as may be read off from the scale provided.

A value of a figure of merit such as this is that one can determine whether one's software is efficient or whether it contains room for improvement.

Powers of 2

The subroutine for $M(I) = 2^I$ taken from Appendix 1 is:

```
2000  !  Subroutine:  Get powers of 2
2010  DIM M(P)
2020  I=0
2030  M(0)=1
2040  M(I+1)=M(I)+M(I)
2050  I=I+1
2060  IF I<P THEN GOTO 2040
2070  RETURN
```

The quantity $P = \log_2 N$, which will have been declared in advance, must be given its numerical value in the dimension statement; for example, if P is 10 then the dimension statement will read `2010 DIM M(10)`. In line 2060 however the symbol P may be used. Then this subroutine assigns values to $M(0)$, $M(1)$, ... $M(P)$. For example, if $P = 5$, the following values will be preassigned to the array $M(\ )$.

I	$M(I)$
0	1
1	2
2	4
3	8
4	16
5	32

The first entry in this table, $M(0) = 1$, is stated in line 2010 and successive entries are obtained by doubling. The doubling is done by addition rather than by multiplication by 2 because the addition is faster than multiplication. After each assignment, execution returns to the doubling operation until the loop has been traversed P times. The time taken should be proportional to P. Therefore, after normalization with respect to NP the vertical height of the strip associated with powers of 2 should die out as N^{-1}. As can be seen from the stripe diagram, the pretabulation of powers becomes an absolutely negligible component of the running time for long data sequences.

Trigonometric functions

A data sequence of length N will require sines and cosines of all integral multiples of one Nth of a turn. Therefore, once N is stated, it pays to set up a precalculated table of those values. One way of doing this, which appears in FHTBAS of Appendix 1, is:

```
3000  !  Subroutine:  Get sines and cosines
3010  W=2*PI/N
3020  A=0
3030  FOR I=1 TO N
3040      A=A+W
3050      S(I)=SIN(A)
3060      C(I)=COS(A)
3070  NEXT I
3080  RETURN
```

As a result of having these values once and for all, any repetitive recomputation is avoided. However, the stripe diagram shows that a good fraction of the whole time goes into getting the trigonometric functions; the values to be calculated are much more numerous than the powers of 2, being proportional in number to N rather than to P. After normalization with respect to NP the corresponding stripe dies away as P^{-1}, but does not die out; therefore considerable interest attaches to reducing the time taken. Much may depend on the computer itself because trigonometric functions are provided internally by a variety of methods according to precision offered and other considerations. One of the best methods is to calculate the tangent t first and then to derive the cosine and sine from $(1+t^2)^{-1/2}$

and $t(1+t^2)^{-1/2}$ respectively. Of course this is not necessary for the user to know and it may come as a surprise to discover by actual measurement that the sine and cosine together can be obtained substantially faster by substituting the following lines for those given above:

```
3030  FOR I=1 TO N/4
3040        A=A+W
3050        C(I)=COS(A)
3060        S(I)=TAN(A)*C(I)
3070  NEXT I
```

Much faster results again are attainable by a fundamental method that goes back to Claudius Ptolemy. The method does not depend upon built-in trigonometric functions at all. A method of this kind is presented in Appendix 1 as part of a subroutine **FHTSUB** that may be incorporated in general programs.

Knowing $\cos(2\pi n/N)$ and $\sin(2\pi n/N)$ for integers n from 0 to $N-1$ is equivalent to knowing the coordinates of N points distributed evenly on a unit circle as in Fig. 8.5(a). But if the coordinates of the points in only one octant are known (Fig. 8.5(b)) the coordinates of all the remaining points can be deduced. Alternatively, one coordinate only of the points in one quadrant will suffice (Fig. 8.5(c)).

Fast sine functions

Tabulation of sines of submultiples of a turn is not a modern endeavor and had reached an advanced level 1800 years ago in the hands of Ptolemy. To compute sines more efficiently than by using the built-in routines supplied by the computer manufacturers, one can learn from previous practice.

Ptolemy's idea was to advance from coarse to fine subdivisions of the circle. From the known chords of 60° and 72° given by the theory of regular polygons one has $\mathrm{ch}\,60° = 1$ and $\mathrm{ch}\,72° = 1.17557$, where $\mathrm{ch}\,\theta = 2\sin\frac{1}{2}\theta$ represents the chord of the angle θ in a circle of unit radius. According to a rather beautiful theorem due to Ptolemy, the chords of differences of given angles can then be obtained from

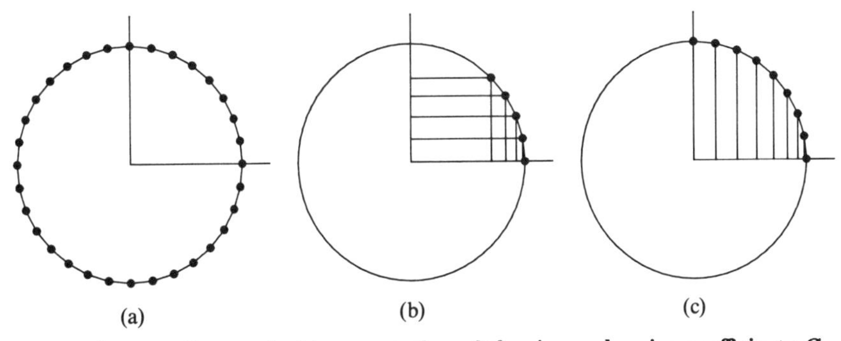

Fig. 8.5. Geometrical interpretation of the sine and cosine coefficients $C_{n,s}$ and $S_{n,s}$ for $s=16$.

$$\mathrm{ch}(\alpha - \beta) = \mathrm{ch}\,\alpha\,\mathrm{ch}(\pi - \beta) - \mathrm{ch}(\pi - \alpha)\,\mathrm{ch}\,\beta,$$

which is equivalent to the more familiar

$$\sin(\alpha - \beta) = \sin\alpha\cos\beta - \cos\alpha\sin\beta.$$

From the known ch 60° and ch 72° he could thus deduce ch 12° and the chords of multiples of 12°. The resultant table is equivalent to a table of sines of multiples of 6°.

Ptolemy's theorem states that, for a quadrilateral $ABCD$ of any shape (Fig. 8.6), whose vertices lie on a circle of center O, $AC.BD = AB.CD + AD.BC$, a discovery that modern students would be hard pushed to derive. When the quadrilateral is a rectangle, Pythagoras' theorem drops out as a special case. With $2\alpha = D\hat{O}B$ and $2\beta = C\hat{O}B$, the difference theorem stated above follows as a trigonometric expression of Ptolemy's theorem.

The table at 12° corresponds to $N = 60$ in the notation of this book. To advance to finer subdivisions Ptolemy introduced a method equivalent to the half-chord formula

$$\mathrm{ch}\,\tfrac{1}{2}\theta = \sqrt{2 - \mathrm{ch}(\pi - \theta)},$$

which yielded results down to 0.75° and multiples thereof, corresponding to $N = 480$. This approach did not lead to results for multiples of 1°. However, Ptolemy argued from an inequality

$$\frac{\mathrm{ch}\,\theta}{\mathrm{ch}\,\phi} < \frac{\theta}{\phi} \quad \text{if} \quad \theta > \phi.$$

Thus

$$\frac{\mathrm{ch}\,1^\circ}{\mathrm{ch}\,0.75^\circ} < \frac{4}{3} \quad \text{and} \quad \frac{\mathrm{ch}\,1.5^\circ}{\mathrm{ch}\,1^\circ} < \frac{3}{2}.$$

Knowing ch 1.5°(= 0.026179) and ch 0.75°(= 0.013090), one may thus conclude that

$$\mathrm{ch}\,1^\circ < 0.0174532 \quad \text{and} \quad \mathrm{ch}\,1^\circ > 0.0174528.$$

Since the two limits agree in the sixth decimal place the argument given is more than a rough approximation and suggests that one might gain by studying this author's methods.[1]

This interesting deduction furnishes the chord of 1° (or sin 0.5°); and

[1] High standards for this tradition had been set by Archimedes who established for example that $265/153 < \sqrt{3} < 1351/780$ (i.e. $1.732026 < 1.732050 < 1.732051$).

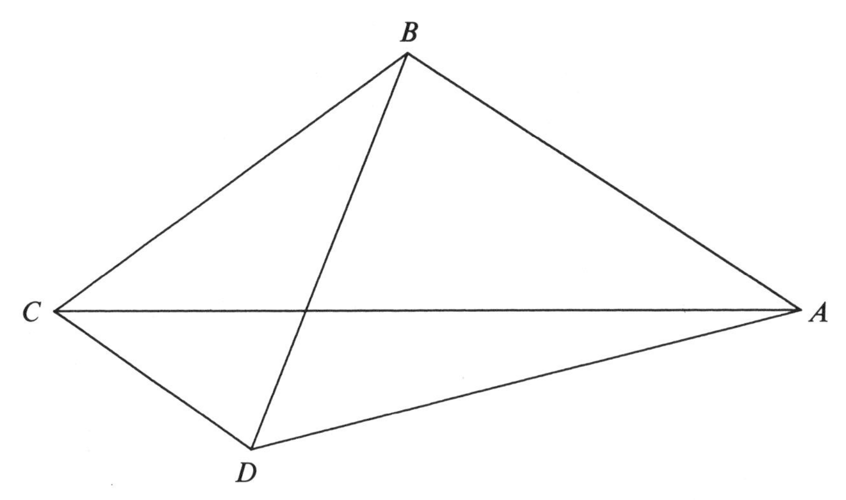

Fig. 8.6. Ptolemy's theorem: $AC.BD = AB.CD + AD.BC$. Can you prove it?

with further application of the halving rule it permitted Ptolemy to set up a table of chords at 0.5° intervals (sines at 0.25° intervals) correct to five decimals. Since the table corresponds to $N = 1440$ it would be adequate for most FFT programs that are run today.

If one starts with a table of $\sin(i\pi/8)$, $i = 1, 2, 3, 4$, that is for θ at intervals of 22.5°, then one can advance to a table at 11.25° intervals by calculating the sine of each intermediate θ from the established values of $\sin(\theta \pm 11.25°)$. The difference formula tells us that

$$\sin\theta = \tfrac{1}{2}\sec\beta[\sin(\theta+\beta) + \sin(\theta-\beta)].$$

The correction factors $\frac{1}{2}\sec\beta$ may be derived from the half-chord formula, starting from the datum $\frac{1}{2}\sec 11.25° = 0.509795579$. Thus

$$\tfrac{1}{2}\sec\tfrac{1}{2}\beta = \frac{1}{\sqrt{2 + \dfrac{1}{\frac{1}{2}\sec\beta}}}.$$

This recursion formula is adopted in **FHTSUB** of Appendix 1.

How to set up the table of cosines is a separate question. In the next section we see that it is desirable to have a table of $\tan\frac{1}{2}\theta$. With such a table cosines could be obtained from the relation

$$\cos\theta = 1 - \sin\theta\tan\tfrac{1}{2}\theta.$$

However, it becomes unnecessary to precalculate cosines explicitly because the tangents of the half angle can be used in their stead and it is advantageous to do so.

Fast rotation

In the inner loop of **FHTBAS** (lines 6170, 6180) there are two operations of the form

$$x = x' \cos\theta - y' \sin\theta$$
$$y = y' \cos\theta + x' \sin\theta$$

which may be recognized as the standard formulae for rotation of coordinates through an angle θ. Four multiplications and two additions are required and, because this inner loop is traversed very many times, the overall running time is critically dependent on efficiency at this point. A method of fast rotation will now be described that was introduced by O. Buneman ("Inversion of the Helmholtz (or Laplace-Poisson) Operator for Slab Geometry," *J. Computational Phys.*, vol. 12, pp. 124-130, 1973). By a rather clever twist the number of multiplications is reduced to three. The number of additions rises to three but for the many computers that take longer to multiply than to add there is a noticeable reduction in running time.

To understand Buneman's algorithm refer to Fig. 8.7. Given (x', y') one obtains the coordinates (x, y) after rotation through an angle θ by projecting the original coordinates onto the x- and y-axes. Thus

$$x = OR - QR = x' \cos\theta - y' \sin\theta.$$

As an alternative, mark off OS so that $OS = x'$. Then $x = OS - QS = x' - QS$. This is an interesting formulation because the term containing x' enters without a multiplying factor. Now the difference QS is the horizontal projection of PT, which is expressible as the sum of $PU = y'$, one of the data, plus UT, which is simply $x' \tan \frac{1}{2}\theta$. Consequently,

$$x = x' - PT \sin\theta,$$

where $PT = y' + x' \tan \frac{1}{2}\theta$.

Knowing x, we can express y as $QV + VP$. Thus

$$y = x \tan \tfrac{1}{2}\theta + PT.$$

The intermediate line segment PT is the catalyst that mediates the result and is represented by T_9 in what follows. The algorithm replacing the standard two-equation transformation is thus:

$$T_9 = y' + x' \tan \tfrac{1}{2}\theta$$
$$x = x' - T_9 \sin\theta$$
$$y = x \tan \tfrac{1}{2}\theta + T_9.$$

This method, which is used in **FHTSUB**, requires a table of $\tan \frac{1}{2}\theta$ and, as mentioned in the preceding section, such a table can also serve the purpose of supplying needed cosines.

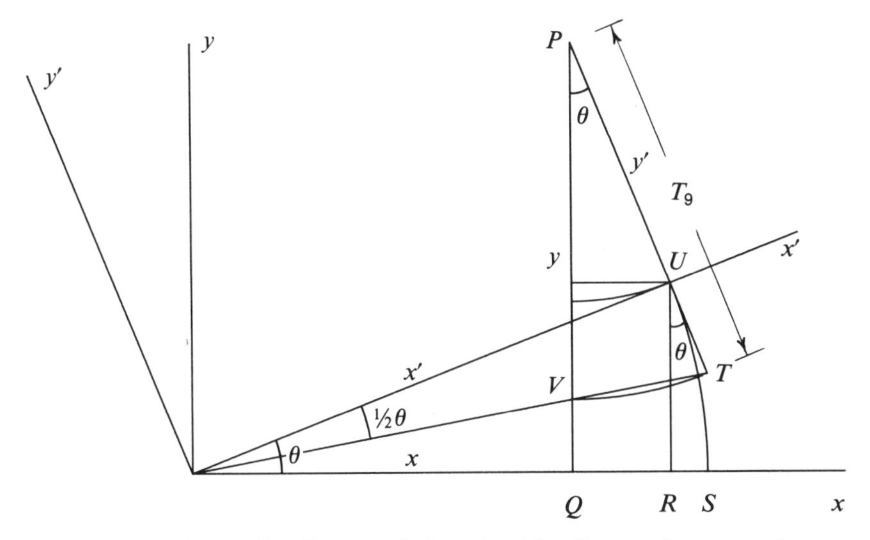

Fig. 8.7. Understanding Buneman's fast algorithm for coordinate rotation.

In order to calculate the tangents we draw on the sine table through the relation

$$\tan \tfrac{1}{2}\theta = \frac{1 - \sin(\tfrac{1}{2}\pi - \theta)}{\sin \theta}.$$

The short segment that does this is

```
9000 K=N4
9000 FOR I=1 TO N4
9000     T(I)=(1-S(K))/S(I)
9000     K=K-1
9000 NEXT I
```

where $S(\ )$ is the sine function and $T(\ )$ is the tangent of half angle.

Fast permutation

As pointed out in connection with the stripe diagram for timing analysis, permutation may occupy a significant fraction of the running time of a general spectral analysis program. The slow or traditional method of permutation utilizes bit reversal (described in Chapter 7) and requires a number of operations proportional to $N(P-1)$. Consequently, even for long data sequences, the fraction of time associated with permutation never dies out. A strategy for permuting more efficiently may be based on the crystal structure pointed out earlier on permutation diagrams. First one determines the coordinates of the cells from

$$x_{\mu,\nu} = N_0\mu + P_{N_0}(\nu)$$

$$y_{\mu,\nu} = N_0\nu + P_{N_0}(\mu),$$

where N_0, the number of cells per side, is given by $N_0 = \sqrt{N/(4 \text{ or } 8)}$ according as P is even (family I) or odd (family II), and $P_N(i)$ is the permutation function for N elements.

Since the total number of cells N_0^2 is proportional to N, the time taken in obtaining the cell coordinates might be proportional merely to N. If this is to be achieved, no undue contribution should originate in the terms $P_{N_0}(\,)$. An asymptotic complexity argument, where operations are counted in the limit of large N, says that since permutation entails $P - 1$ stages of N operations each, then the computation time for $x_{\mu,\nu}$ will approach proportionality to $N(P - 1)$. However, an interesting way has been found of circumventing this consequence. Because N_0 becomes smaller and smaller relative to N as N increases, the time spent on $P_{N_0}(i)$ becomes insignificant. The successive additions needed to establish $x_{\mu,\nu}$ and $y_{\mu,\nu}$ dominate the time.

After the cell coordinates are calculated there are always only 4 or 8 exchanges to be made because the number of atoms per cell does not increase with increasing N but only alternates. Timing experiments confirm that the geometrical approach requires a time proportional to N. The vertical thickness of the stripe on the timing diagram, where time is normalized with respect to NP, therefore declines as P^{-1} and becomes negligible.

The method described here is illustrated in **FASTPERMUTE**, a demonstration program that is presented for study in Appendix 1. Imbedded in the program is a permutation program to obtain $P_{N_0}(i)$; in the olden days we would have called this definitio in circulo; now we call it bootstrapping. Paradoxically the program as a whole runs faster than the one it contains. A recursive formulation can be imagined that permutes the sequence of reduced length by reducing it again. Very soon the length would be reduced to 2, whereupon no further permutation would be required.

Radix 4 and other refinements

For purposes of exposition attention has been restricted to sequences of length N where N is equal to 2, 4, 8, 16, ... ; in short where N is expressed as radix 2 raised to a power P:

$$N = 2^P.$$

But what if one has 300 items of data? Possibly 44 items could be abandoned to reduce the number to 256, perhaps the data gathering procedure could be adjusted by the next occasion, or perhaps 212 trailing zeros could be added to bring the length to 512. It is clearly wasteful of computing time to handle artifically lengthened sequences; as a result, data gathering instruments are normally designed to harmonize with 2^P and the computation and display of digital images is planned for 512 x 512 or other sizes of the form $2^{P_1} \times 2^{P_2}$.

Some instrument design is, however, out of our hands; for example the number of lines in a broadcast television image, which is between 512 and 1024 and in some cases as low as 525. Also geophysical time series, astronomical data and other data of natural origin have features not under control. The sunspot number series has 286 terms in 1985 because the series starts in 1700 and there is not much we can do about N except wait.

If we go back to the reason why 2^P arose we recall that it was because of a principle of divide and conquer whereby lengths were reduced stage by stage until length 2 was reached. Then the transforms of length 2 were trivial and all the 2-element transforms were combined. Clearly the 3-element transform

$$\tfrac{1}{3}\sum_{\tau=0}^{2} f(\tau)\,\mathrm{cas}(2\pi\nu\tau/3)$$

is also trivial: its butterfly diagram is shown in Fig. 8.8. As Gauss explained, any factors that the length possesses will permit the work of computation to be reduced.

Of course some extra multiplications are required when the factors are not all 2 but there is a net saving. The algorithm that carries this idea to the limit for the DFT by allowing N to be expressed in the form

$$N = 2^{P_1}3^{P_2}5^{P_3}7^{P_4}\ldots,$$

where the successive radices are the prime number series, was perfected by S. Winograd. For an explanation and further references see C.S. Burrus and T.W. Parks, *DFT/FFT Convolution Algorithms* (Wiley Interscience, 1985). Permutation algorithms for radices other than 2 are given by these authors.

For each user with a refractory N there are, however, many for whom it is possible to tailor their N to accommodate whatever N is most favorable as regards speed. If we ask about N of the form 4^P, which is even

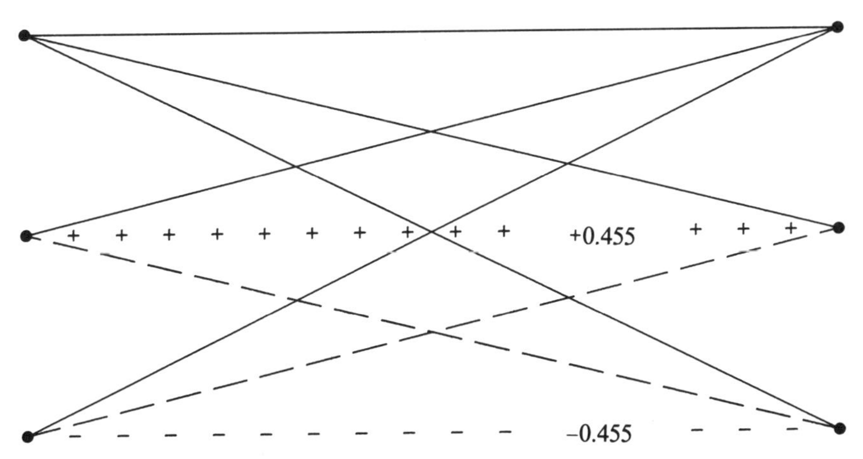

Fig. 8.8. Butterfly diagram for a 3-element DHT. The full lines represent a transmission factor $\frac{1}{3}$, the broken lines a factor $(\sqrt{3}-1)/6 = 0.122$, and the remaining lines $\pm(\sqrt{3}-1)/6 = \pm 0.455$.

more restrictive than 2^P, this latter class of user will find a desirable speed advantage. In the 4-element DHT, the transmission factors are either 1 or -1 and thus do not entail multiplication and in the 4-element butterflies required when $N = 4^P$ and reduction stops at 4 elements, the total number of multiplications is noticeably reduced, typically by about one-third. Consequently the radix-4 transform is a significant refinement. A program FHTRX4 is presented in Appendix 1.

Earlier approaches to real data

When data are real rather than complex, which is usually the case, the FFT is an inefficient method of transforming, because it provides for complex input. On the other hand, once we are in the transform domain manipulating complex coefficients and wish to return to the domain of real data, the FFT is perfectly equipped to take the complex input. But to call on the FFT merely to supply real output when it has the capacity for complex output, is to call on a program that is overqualified for the task and therefore again inefficient, though in a different way.

A variety of approaches may be taken to improve on the FFT for spectral analysis of real data. If one is planning to visit the transform domain and stay there, as is the case with the computation and display of power spectra, a function performed by commercial spectrometers for example, then a way to speed up the FFT is to strike out the variables representing the nonexistent imaginary input. This stripped down algorithm is of course no longer a Fourier transform (for one thing it lacks the reciprocal property) but, all things being equal, ought to run in the same time as the FHT.

If one is planning to visit the transform domain, perform operations such as filtering, convolution or other signal processing, and then return to the domain of real data, the stripped down algorithm is not available because the input is now complex. However, a second specialized algorithm can be constructed that accepts complex input and provides only for real output. Consequently by storing two special algorithms, each of which is an appropriate modification of the FFT, one can achieve a speed of comutation that is comparable with that of the FHT. This approach has been adopted in commercial software.

By comparison with the FHT, the procedure described requires storage space for two different transforms. The user must also keep track of which one to use, and array storage must be organized for real and imaginary variables because both transforms involve complex arithmetic. While the results are satisfactory for users of canned software, yet from the standpoint of one wishing to modify or maintain such a program or to incorporate it into larger programs, the code is awkward as well as bulky. Other approaches, however, have been utilized.

A sequence of real data

$$\{a \quad b \quad c \quad d \quad e \quad f \quad g \quad h\}$$

may be compacted into the shorter complex sequence

$$\{a+ib \quad c+id \quad e+if \quad g+ih\}$$

The FFT may be applied to the complex sequence to obtain $\{\alpha+i\beta \quad \gamma+i\delta \quad \epsilon+i\varsigma \quad \eta+i\theta\}$. The even and odd parts of the sequences $\{\alpha \quad \gamma \quad \epsilon \quad \eta\}$ and $\{\beta \quad \delta \quad \varsigma \quad \theta\}$ can now be calculated and combined with appropriate factors, as called for by the shift theorem, to emerge with the desired FFT

$$\{A \quad B+iB' \quad C+iC' \quad D+iD' \quad E \quad D+iD' \quad C-iC' \quad B-iB'\}.$$

From this description it is clear that the speed improvement comes from the use of a 4-element FFT rather than the inefficient 8-element FFT. To return from the transform domain the same economy can be achieved. In this case, one truncates the 8 elements to 4 elements, which clearly can suffice to specify the 8-element transform. All but one of the remaining coefficients may be assumed to be complex conjugates of ones that are retained. The fifth element E may be assumed to be real; therefore it can be tucked in with the first element A which is also always real. The truncated representation is $\{A+iE \quad B+iB' \quad C+iC' \quad D+iD'\}$. After transformation using the inverse FFT one obtains a 4-element complex sequence out of which the desired 8-element real sequence can be conjured with some further manipulation. Suitable programs have been clearly documented in a standard reference (*Programs for Digital Signal Processing*, IEEE Press, 1979).

These ingenious forerunners are made obsolete by the Hartley formalism which cleanly avoids complex arithmetic and which allows one and the same program to be applied equally whether visiting or returning from the transform domain. Where complex numbers are really required they are simply constructed as a final step. Interestingly, real and imaginary parts are needed in their own right in only a fraction of problems. They are commonly used as intermediate stages toward a final goal because thought patterns have benefited from the convenience of complex algebra. Thus a power spectrum is often thought of as the sum of the squares of the real and imaginary parts. But if one were computing with real data by real arithmetic it would be an unnecessary digression to go to real and imaginary parts in order to get the power spectrum because it can equally well be reached without help of the complex plane at all. Thus, starting from the DHT $H(\nu)$ we get the power spectrum $P(\nu)$ directly from

$$P(\nu) = [H(\nu)]^2 + [H(N-\nu)]^2.$$

The same applies to phase $\phi(\nu)$ which is commonly thought of in terms of the ratio of the imaginary and real parts but may equally be expressed in terms of the even and odd parts of the purely real Hartley transform. Thus

$$\phi(\nu) = \arctan\left(\frac{H(\nu)+H(N-\nu)}{H(\nu)-H(N-\nu)}\right).$$

Problems

8.1 *Secant formula.* Derive the relation $\frac{1}{2}\sec\frac{1}{2}\beta = (2+2/\sec\beta)^{-\frac{1}{2}}$ from the half-chord formula $\mathrm{ch}\frac{1}{2}\theta = \sqrt{2-\mathrm{ch}(\pi-\theta)}$.

8.2 *Fast secant formula.* (a) Show that the square root of a quantity x that is near unity is given to high accuracy by $\sqrt{x} = (2a+1)/(2a-1)$ where $a = (x+1)/(x-1)$. (b) Apply this rule to $x = 2$, which is only a rough approximation to unity, to show that accuracy of one per cent is achieved. (c) Show that the correction factor $\frac{1}{2}\sec\frac{1}{2}\beta$ is obtainable to eight-figure accuracy, for $\beta \le 11.25°$, from Buneman's secant approximation

$$\tfrac{1}{2}\sec\tfrac{1}{2}\beta = 0.7001 - \frac{0.16016004}{0.3004 + \frac{1}{2}\sec\beta}.$$

8.3 *Rotation identity.* Show how the identity

$$\tan\tfrac{1}{2}\theta = \frac{(1-\cos\theta)}{\sin\theta}$$

may be interpreted geometrically.

8.4 *Buneman's rotation algorithm.* Deduce Buneman's algorithm directly from the coordinate transformation $x = x'\cos\theta - y'\sin\theta$.

8.5. *Ptolemy's theorem.* Prove Ptolemy's theorem.

8.6. *Difference formula.* Deduce the trigonometrical formula $\sin(\alpha-\beta) = \sin\alpha\cos\beta - \cos\alpha\sin\beta$ from Ptolemy's theorem.

8.7 *Trisection.* Use the identity for $\mathrm{ch}(\alpha+\beta)$ in the form $\mathrm{ch}(2\alpha+\alpha)$ to show that $\mathrm{ch}\,\frac{1}{2}°$ could have been deduced from $\mathrm{ch}\,1\frac{1}{2}°$ by the formula $\mathrm{ch}\,\frac{1}{2}° = \frac{1}{3}\mathrm{ch}\,1\frac{1}{2}° + (4/81)\,\mathrm{ch}^3\,1\frac{1}{2}°$.

8.8. *Inverse permutation.* What is the inverse matrix of the 8×8 permutation matrix $\mathbf{P}_8$?

8.9 *Place notation.* Find out how Claudius Ptolemy could present trigonometric tables accurate to five decimals more than a millennium before the decimal system came into use.

CHAPTER 9

AN OPTICAL HARTLEY TRANSFORMER

"If you destroy Logopolis you unravel the whole causal nexus."

Dr. Who

Images may be processed digitally in an electronic computer but there are also analogue methods. Indeed, before an image of an external object even reaches a computer it will have been operated upon by analogue optical systems which may perform such signal processing functions as low pass filtering and sharpening. Normally these optical operations are discussed in terms of the two-dimensional Fourier transform, but since we now know that for every application of the Fourier transform there is a corresponding application of the Hartley transform, it is of interest to inquire whether the two-dimensional Hartley transform may have any relevance to optical processing. In this chapter we shall show that indeed it does.

As an historical comment it may be noted that although Hartley's transform was published in connection with telephone transmission systems, Hartley did other work that has had a connection with optics. In fact Gabor's concept of the metrical information in light beams [D. Gabor, "Light and Information," in *Progress in Optics*, E. Wolf, ed., vol. 1, pp. 109-153 (North-Holland, 1961)] traces directly back to "Hartley's Law," which encapsulated the discovery that the quantity of information transmitted by a telephone line is proportional to the product of the bandwidth and the duration of the transmission. Seemingly obvious today, this law laid the basis for "Shannon's Law," which added the idea of symbol probability, and led ultimately to Gabor's measure for the quantity of information in a light beam.

Optical relevance

Apart from the general application to computing it is not immediately clear that the Hartley transform would be relevant to analogue devices. After all, it is the Fourier transform which is inherent in so many of the phenomena of nature, for example the operation of a lens with respect to its two focal planes, not some other transform. A coherently illuminated object in the front focal plane produces the *Fourier* transform in the back focal plane, not the Hartley transform. Thus, for any real input, the image in the back focal plane possesses twofold conjugate rotational symmetry (the field at any point is the complex conjugate of the field at the point on the opposite side of the optical axis at the same distance). The Hartley transform in general does not possess this degeneracy.

It is true that a lens does not, strictly speaking, in fact produce the complex Fourier transform, because complex numbers are a figment of the human mind rather than properties of the physical world. Speaking carefully, one would have to admit that the fields in the so-called Fourier plane are actually real. In any case, when we use complex numbers to represent the actual situation, it is the Fourier transform to which we are led for the mathematical representation of the physical state.

Despite the appositeness of the Fourier transform for describing the operation of a lens there are nevertheless some features of the Fourier transform that are inconvenient. For example, the phase of the field contains substantial information but available optical sensing elements respond only to intensity and not to phase. A photographic record made in the Fourier transform plane thus abandons phase information, which can be very important. On the other hand, direct recording of the squared modulus of the Hartley transform would provide much richer information. In cases where the transform does not go negative $|H(u,v)|^2$ suffices to recover $f(x,y)$ in full; in other cases there are regions of the (u,v)-plane within which the sign of $H(u,v)$ is negative and knowledge of $|H(u,v)|^2$ does not of itself determine the sign. However, sign ambiguity is a much less serious defect than absence of phase knowledge and can often be resolved simply.

Since access to a real field offers advantages it is desirable to ask how to design an optical system that presents the Hartley transform of a given object.

Symmetry considerations

The Hartley transform $H(u,v)$ is the difference of the real and imaginary parts of the Fourier transform $F(u,v)$. Let

$$F(u,v) = F_{real}(u,v) + iF_{imag}(u,v).$$

Then

$$H(u,v) = F_{real}(u,v) - F_{imag}(u,v),$$

where

$$F_{real}(u,v) = \tfrac{1}{2}[F(u,v) + F(-u,-v)]$$

and

$$iF_{imag}(u,v) = \tfrac{1}{2}[F(u,v) - F(-u,-v)].$$

First we ask how to form $F_{real}(u,v)$. One answer to this would be to split the field in the Fourier transform plane into equal parts, to rotate one component through an angle π about the optical axis so as to obtain $F(-u,-v)$, and to recombine.

This can be done with a Michelson interferometer in one arm of which there is a cube corner and in the other a plane mirror at the same distance as the vertex of the cube corner (Fig. 9.1). Alternatively, a roof prism may be placed in each arm, the ridgelines forming an orthogonal system with the optical axis (Fig. 9.2). One roof prism inverts the field about an axis parallel to the ridgeline and the other prism inverts about the perpendicular axis to

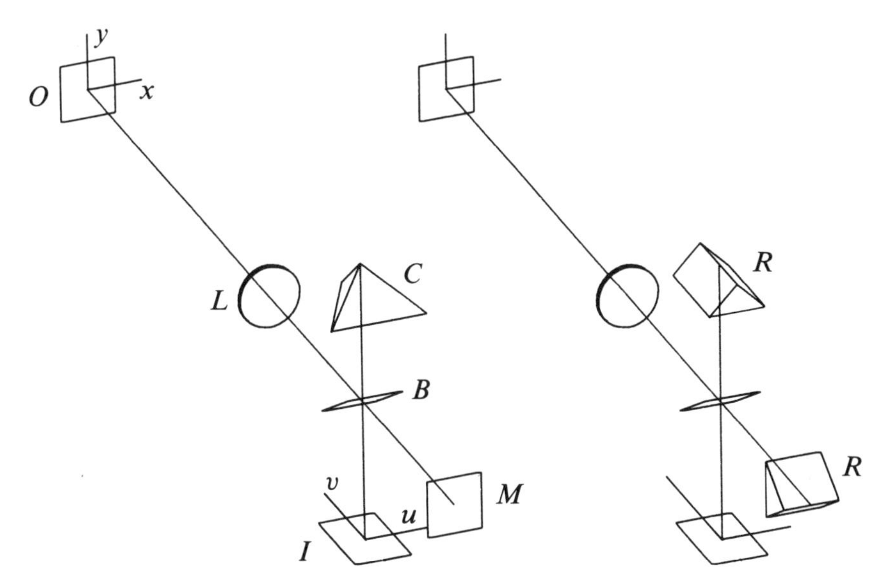

Fig. 9.1. (Left) A Michelson-type interferometer in which one beam is rotated half a turn by a cube corner and superimposed on itself. Object plane is O, lens L, beamsplitter B, plane mirror M with quarter wave plate, image plane I, analyzer A.

Fig. 9.2. (Right) A roof-prism interferometer that also recombines the components of a split beam with half a turn of relative rotation.

produce the desired rotation by π. The cube corner produces this rotation entirely in the one arm with three reflections. The resulting combined image will be reversed as compared with the roof prism arrangement, an attribute that may be a design consideration. A third optical arrangement, shown in Fig. 9.3, may be based on the image rotation that takes place in a telescope.

To form $F_{imag}(u,v)$ the same arrangements may be used except that sign reversal must be introduced in one arm. A monochromatic method for doing this is to produce a half-wave delay by inserting phase plates into each arm differing in phase delay by $\pi/2$; this will result in sign reversal by introducing a net relative phase of π. Alternatively a single coating of phase thickness $\pi/2$ could be applied to the surface of the reflecting device in one arm only. As another alternative, the physical position of one of the reflectors could be adjusted by $\lambda/4$.

Implementation using polarization

The discussion above describes how to realize either F_{real} or F_{imag} using separate optical arrangements. In order to have both F_{real} and F_{imag} simultaneously present for subtraction one may illuminate the object with linearly polarized light in a direction midway between the x- and y-axes. The horizontal and vertical components of polarization would then pass through the equal-armed roof prism system identically and produce the

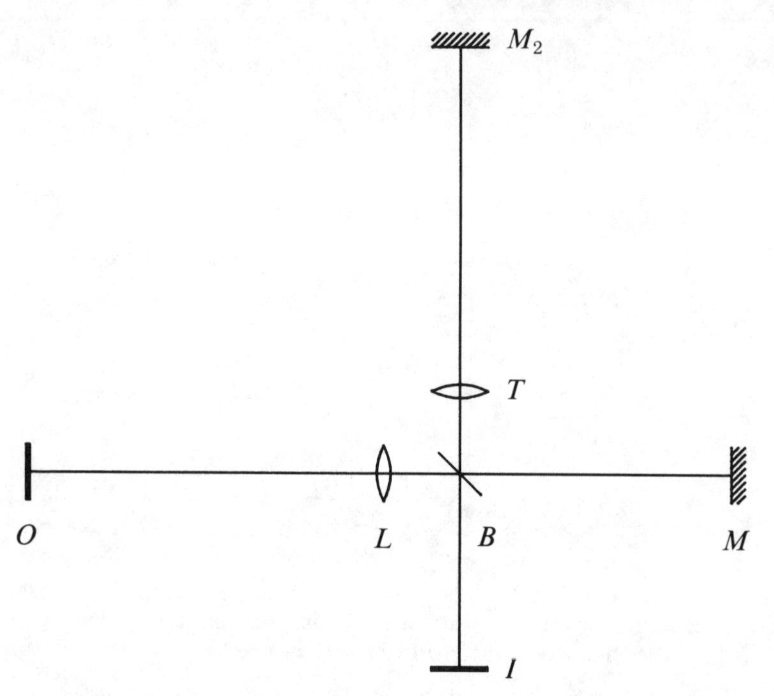

Fig. 9.3. An interferometer that uses the rotation that occurs in a telescope.

same F_{real} as before. Now if a suitably oriented quarter-wave plate is inserted in one arm there will be an excess path length of half a wavelength introduced between the two polarizations (after two transits) so that F_{real} will be produced at the output plane in one polarization and F_{imag} in the other polarization. The net vibration in the direction midway between the u- and v-axes will then be either the Hartley transform $F_{real} - F_{imag}$ or, depending on conventions, $F_{real} + F_{imag}$, which is the same thing rotated through half a turn in the (u, v)-plane. It may be viewed by a suitably oriented analyzer.

Originally conceived in this way, the Hartley transformer has a simpler implementation described by R.N. Bracewell, H. Bartelt, A.W. Lohmann, and N. Streibl, "Optical Synthesis of Hartley Transform," *Applied Optics*, vol. 24, pp. 1401-1402, 1985.

A practical implementation

Instead of regarding the two-dimensional Hartley transform as made up of the difference of the real and imaginary parts of the Fourier transform, one may manipulate the definition in another way that leads to a simplified optical arrangement, as follows. Remembering that

$$\sqrt{2}\cos\left(x - \frac{\pi}{4}\right) = \cos x + \sin x = \operatorname{cas} x,$$

one may write the Hartley transform in the form

$$H(u,v)=\sqrt{2}\int_{-\infty}^{\infty}\int_{-\infty}^{\infty}f(x,y)\cos\Big[2\pi(ux+vy)-\frac{\pi}{4}\Big]dxdy.$$

Starting from this basis one sees that the Hartley transform can be synthesized from two phase-adjusted Fourier transforms by the following development:

$$\begin{aligned}H(u,v)&=\sqrt{2}\int_{-\infty}^{\infty}\int_{-\infty}^{\infty}f(x,y)\tfrac{1}{2}[e^{2\pi i(ux+vy)-i\pi/4}+e^{-2\pi i(ux+vy)+i\pi/4}]dxdy\\&=\frac{1}{\sqrt{2}}e^{-i\pi/4}F(-u,-v)+\frac{1}{\sqrt{2}}e^{i\pi/4}F(u,v)\\&=\frac{e^{i\pi/4}}{\sqrt{2}}[F(u,v)+e^{-i\pi/2}F(-u,-v)].\end{aligned}$$

Now the constant phase factor $e^{i\pi/4}$ is only a matter of the choice of time origin. Consequently the combination of Fourier transforms indicated can be arrived at by means of a Michelson-type interferometer in one arm of which there is a cube corner that rotates the field $F(u,v)$ through an angle π to produce $F(-u,-v)$. In the other arm one may place a plane mirror at the same distance from the beamsplitter as is the vertex of the cube corner just as in Fig. 9.1. Alternatively, the double roof prism arrangement of Fig. 9.2 or the telescope arrangement of Fig. 9.4 may be adopted, and other configurations making use of Dove prisms or a Mach-Zehnder-type interferometer can be imagined.

After the wave fields have been brought together with the relative rotation π, a phase factor $e^{-i\pi/2}$ has to be introduced into one arm of the interferometer. One way to do this is to incorporate a phase plate and another way is to shift one of the mirrors by an eighth of a wavelength. Neither of these adjustments affects the appearance of the illustrations but the effect produced is profoundly different. If the source in the object plane is coherent this arrangement produces the desired Hartley transform and a photographic plate placed there would record the squared modulus of the Hartley transform.

Incoherent object

Coherent sources are not the only ones that arise and it is interesting that the Hartley transform of an incoherent source can also be generated, but in this case as an intensity distribution superimposed on a steady bias. Consider an element of an incoherent source located at (x_0,y_0) in the (x,y)-plane. This element produces in the transform plane an interference pattern whose intensity is proportional to $1+\cos[4\pi(x_0u+y_0v)-\phi]$, where ϕ denotes the difference in phase length between the two arms of the interferometer. Since intensities are additive under incoherent conditions, all the interference patterns for the various source elements may be added together to produce a total intensity I_{out} that is given by the integral

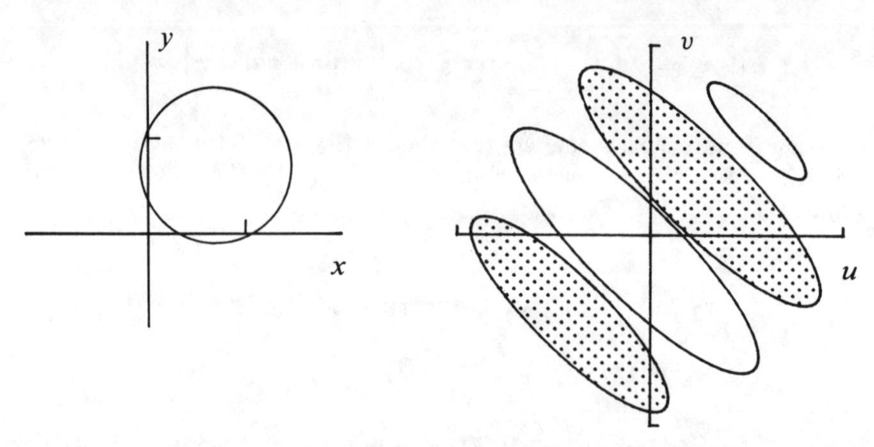

Fig. 9.4. A transparency (left) and the amplitude distribution in the Hartley plane (right) as represented by the contours at ±0.125 times the peak. Only four regions are intersected by these two contour levels.

$$I_{out} = \int_{-\infty}^{\infty}\int_{-\infty}^{\infty} I_{in}(x,y)dxdy + \int_{-\infty}^{\infty}\int_{-\infty}^{\infty} I_{in}(x,y)\cos[4\pi(ux+vy)-\phi]dxdy.$$

By a proper choice of ϕ we again obtain the Hartley transform, this time of the object intensity distribution, superimposed on a steady bias. An interesting feature of the bias is that it automatically removes any ambiguity as to the sign of the Hartley transform.

The Hartley transformer could be cascaded for signal processing purposes. As an alternative, by the use of a mirror in the transform plane, the Hartley transformer could also be used in the reverse direction. A filter in the transform plane, either in transmission or reflection, would permit convolution to be performed in an analogue fashion in those cases where the convolving factor is an even function.

Example

In Fig. 9.4 a transparency with a finite Gaussian spot at (X, Y) in the first quadrant is shown. The transmissivity of the object could be represented mathematically by

$$f(x,y) = \exp\{-\pi[(x-X)^2/A^2 + (y-Y)^2/B^2]\},$$

where the equivalent widths are A and B in the x- and y-directions respectively. This spot is represented by a single contour at the +0.125 level. The Hartley transform, for reference, is

$$H(u,v) = AB\operatorname{cas}[-2\pi(Xu+Yv)]\exp[-\pi(A^2u^2+B^2v^2)]$$

and the Fourier transform is

$$F(u,v) = AB \exp[-i2\pi(Xu+Yv)] \exp[-\pi(A^2u^2+B^2v^2)].$$

For the real and imaginary parts we have

$$F_{real}(u,v) = AB \cos[2\pi(Xu+Yv)] \exp[-\pi(A^2u^2+B^2v^2)]$$

$$F_{imag}(u,v) = -AB \sin[2\pi(Xu+Yv)] \exp[-\pi(A^2u^2+B^2v^2)].$$

The Hartley transform $H(u,v)$ is represented in the (u,v)-plane by contours at levels $\pm AB/8$. The phase of the field would be the same all over the (u,v)-plane (in the polarization selected by the viewing analyzer) save in the shaded patches where the amplitude is negative. By contrast the phase of the Fourier transform progresses continuously by almost three full turns across the same field.

In general, however, null contour lines on the (u,v)-plane separate positive and negative regions of the Hartley transform so that the sign ambiguity existing in an intensity record such as a photograph would often be a trivial concern. It is true that objects symmetrical with respect to the optical axis may in special cases give rise to null loci across which the sign does not reverse, but the resulting ambiguity can be dispelled by breaking the symmetry by a shift of the object. However, the sign is easy to determine. One way of identifying the negative regions would be to open a small aperture at $x=0$, $y=0$ so as to let through a weak isophase reference wave; the intensity would increase in a positive region but a negative region would reveal itself by a drop in intensity. A weak time-modulated isophase reference beam, e.g. a chopped beam, would reveal the regions of negative amplitude through the time modulation produced in the transform plane; in a negative-amplitude region the time modulation of intensity would be in antiphase with that in a positive region. A steady reference beam could also be switched on temporarily for the purpose of an auxiliary sign-determining exposure. Cinematography of moving objects could make use of a reference beam that alternated between successive exposures.

Interestingly, superposition of a zero-phase reference beam originating at $x=0$, $y=0$ can be used as more than a diagnostic tool for sign determination; one could also incorporate such a beam in special cases to lift some or all of the negative regions completely into the positive condition. By this means negative amplitudes, which are an inconvenience in certain kinds of optical signal processing, could be removed.

Holography in the Hartley plane

Of course it is possible to record a complex field by holography. Although holography is rather more complicated than simple photographic recording in the focal plane, holographic technique could nevertheless also be applied to the Hartley image plane. A Hartley-based hologram of a real object would differ from a normal hologram in that the position of the fringes would be controlled mainly by the path length to the source of the reference beam rather than depending on both path length and illumination phase in the transform plane; of course there would be jumps at the boundaries of negative regions. In the special case of an isophase reference beam, such as one originating at the origin of the object plane, the hologram would be particularly simple. The procedure would be the same as the means mentioned in the previous section for lifting the negative regions to positive levels except that the reference beam would now be strong relative to the amplitude range of the Hartley transform, as is characteristic of reference beams in holography. This new type of hologram would be more convenient to make, copy and transport than a hologram composed of fringes with submicron spacing, because the fringes would be much coarser. Fringe spacing would have to do with the structure of the object rather than with the wavelength of the radiation.

Replacing the real object by a complex one in the same optical set-up would result in a more general hologram containing the fine structure found in conventional holograms. However, if a special hologram can be made of a real object the same can be done for an imaginary object; thus there is a way of representing a complex object by a pair of special holograms. To do this would require image rotation and superposition of the kind that has already been described for the Hartley plane so as to generate the real and imaginary parts of the complex object. One of the advantages of dealing with real fields rather than complex ones is that photographic recording, which ignores phase, can capture all the information. The same is true of cathode-ray tube phosphors, photomultipliers, infrared detectors and many other sensors which, unlike radio receivers, do not respond to phase.

APPENDIX 1

COMPUTER PROGRAMS

"And whan that he wel dronken hadde the wyn,
Than wolde he speke no word but Latyn." *Chaucer*

This appendix contains a number of original programs that are designed for general usability and so are liberally sprinkled with explanatory remarks preceded by !. On retyping these programs, the user may omit any line which is entirely a comment or any part of a line from ! on. The programs are in readable BASIC and are intended for users who in general will adapt the programs for their own convenience. This class of user generally finds it easier to go from BASIC to FORTRAN, PASCAL or C than in any other direction.

Line numbers have been tailored to enhance readability, for example by the use of multiples of 1000 for numbering subroutines. Similar subroutines in separate programs may thus be coded by line number. Subprogram calls and their requirement for parameter lists may be avoided in this way but it is presumed that users who make use of the same subroutines frequently will construct their own main programs in terms of **CALL** commands as suits them. In the small number of short programs presented here there is no call for **CALL**. All the programs are clearly structured to consist of a preamble, main program, data where applicable, and subroutines. **GOTO** commands are used only where they are superior to the alternatives and **FOR-NEXT** loops are emphasized by indentation. Algebraic symbols, italicized as is customary, have been used in **DIM** declarations and elsewhere where numerical values are to be substituted.

Sample printouts and notes have been appended to each program and careful attention has been given to the visual presentation to make the programs easy to study.

Local features

A strength of BASIC is that improvements have flowed steadily from the computer industry's pool of language designers competing to meet the needs of students and owners of personal computers. Consequently the language is now very widely understood and, not having been prematurely rigidified by standardization, is up to date. The better features have been copied freely throughout the industry. This dynamic development means that local features need to be flagged. In the programs presented the following items may be noted.

! introduces a remark that is not a command to the computer. REM may be substituted, but the single symbol proves to be very convenient in the course of program development because it can easily be slipped in and out of single spaces.

@ is a statement separator that permits more than one statement under the one line number. Use of **@** conserves space, may be utilized to improve readability, and increases the speed of some operations. Usually the separate statements may be rewritten in sequence under consecutive line numbers. However one of the conveniences of **@** is to handle conditional statements in a natural way without brackets or repetition as in

```
999 IF B > A THEN C = 2 @ D = D/10
```

If the condition **B > A** is not met then **C** is of course not set to 2; but neither is **D** reduced to one-tenth; action passes straight to the next line number.

Variables are **A ... Z, A0 ... Z9**; array variables are **A() ... Z9** and **A(,) ... Z9(,)**; and string variables are **A$ ... Z9$** . String arrays **A$()** do not occur in these programs.

Since spaces are ignored in BASIC, the indentation of **FOR-NEXT** loops is for clarity of presentation only and need not be preserved in copying.

The function **VAL$(X)** is the string expression corresponding to the numerical value of the variable **X**.

Guide to the programs

FHTBAS is a demonstration program for the fast Hartley algorithm that computes the discrete Fourier transform of a series of real numbers and prints the complex result together with the DHT from which the DFT is derived. It is written to be studied in connection with text discussion of the various subroutines.

FHTSUB is a fast in-place algorithm in the form of a subroutine suitable for incorporation into general programs. It delivers the DHT of a real sequence. The program has been condensed, and proceeds sequentially rather than by going to subroutines within the subroutine, but line numbering is keyed to that of **FHTBAS** so that the structure can be retrieved if desired. More sophisticated methods are used for trigonometric functions and for permutation so that running time will be improved. The in-place computation reduces the demand on memory and permits higher values of N to be reached. The methods used for the pretabulated functions are generally applicable and are based on the work of O. Buneman, "Conversion of FFT's to Fast Hartley Transforms," *SIAM J. Sci. Statist. Comp.*, vol. 00, pp. 000-000, 1985. The permutation method is due to David M. Evans. For related programs in FORTRAN see Z. Wang, "Fast Algorithms for the Discrete W Transform and for the Discrete Fourier Transform," *IEEE Trans. Acoustics, Speech and Signal Processing*, vol. ASSP-32, pp. 803-816, 1984, and (for a radix-4 version among others) H.V. Sorensen, D.L. Jones, S.C. Burrus and M.T. Heideman, "On Computing the Discrete Hartley Transform," *IEEE Trans. Acoustics, Speech and Signal Processing*, 1985 (in press).

FHTPS computes the power spectrum of a real sequence by way of the DHT and illustrates the use of **FHTSUB** as a subroutine appended to a special program. Provision is made for discretionary smoothing. The power spectrum is obtained without appeal to the Fourier transform at all.

CONV computes the ordinary convolution $f_1 * f_2$. For the ordinary cross-correlation of f_1 on f_2, reverse f_1.

CCONV computes the cyclic convolution $f_1 \circledast f_2$. For the cyclic cross-correlation $f_1 \circledast f_2$, reverse f_1.

MATCON performs convolution compactly by a matrix multiplication command.

ACF performs autocorrelation.

CACF does cyclic autocorrelation.

ICONV performs inverse convolution, or deconvolution.

RECIP inverts a real sequence to obtain the reciprocal sequence which, when convolved with the given sequence, yields $\{1\}$.

ICORR inverts autocorrelation, yielding an original sequence possessing the autocorrelation that is given.

FHTCONV performs convolution by multiplication in the transform domain.

FHTACF performs autocorrelation by multiplication in the transform domain.

FHTRX4 is a radix-4 version of the DHT.

FASTPERMUTE is a demonstration program that takes a sequence of length N and permutes it by use of a slower permutation program of much shorter length N_0, where N_0 is both the number of cells per side and also the grid spacing within each cell.

FHTBAS.FOR is a FORTRAN version of the demonstration program **FHTBAS** given above in BASIC. It is rather simple to convert the BASIC programs to FORTRAN; this example will serve as a Rosetta stone. The principal differences are **DO** loops for **FOR-NEXT** loops and quaint phraseology such as **(I .LT. P)** for **I < P**.

FHTFOR.FOR is a FORTRAN version of **FHTSUB** written as a subroutine subprogram to be called by a main program.

```
10 !                              "FHTBAS"
20 DIM F(8,n) !  Hartley stages
30 DIM R(½n + 1), X(½n + 1) !  Fourier transform
40 DIM S(¼n), C(¼n) !  sin & cos
50 P = p
60 N4 = 2∧(P - 2) @  N2 = N4 + N4 @  N = N2 + N2 @  N7 =
       N - 1 @ P7 = P - 1
70 DEF FN F(I) = I + 1 !  Sample function

100 TO = TIME ! Start timer
110 GOSUB 1000 !  Insert data
120 GOSUB 2000 !  Get powers of 2
130 GOSUB 3000 !  Get C() & S()
140 IF P > 1 THEN GOSUB 4000 !  Permute
```

```
150 GOSUB 5000 !  Stages 1 & 2
160 GOSUB 6000 !  Stages 3,4....
170 GOSUB 7000 !  Get DFT
180 TO = TIME - TO ! Stop timer
190 GOSUB 8000 !  Print results
200 END

1000 !  Subr Insert data
1010 FOR I = 0 TO N7 @ F(0,I), F(1,I) = FNF(I) @ NEXT I
1020 RETURN

2000 !  Subr Get powers of 2
2010 I = 1 @ M(0) = 1 @ M(1) = 2
2020 M(I + 1) = M(I) + M(I)
2030 I = I + 1
2040 IF I < P THEN GOTO 2020
2050 RETURN

3000 !  Subr Get sines & cosines
3010 W = 2*PI/N @ A = 0
3020 FOR I = 1 TO N4 @ A = A + W @ S(I) = SIN(A) @ C(I) =
      COS(A) @ NEXT I
3030 RETURN

4000 !  Subr Permute
4010 J, I = -1
4020 I = I + 1 @ T = P
4030 T = T - 1 @ J = J - M(T)
4040 IF J > = -1 THEN GOTO 4030
4050 J = J + M(T + 1)
4060 IF I <= J THEN GOTO 4020
4070 T = F(0,I + 1)
4080 F(0,I + 1) = F(0,J + 1)
4090 F(0,J + 1) = T
4100 IF I < N - 3 THEN GOTO 4020
4110 RETURN

5000 !  Subr Stages 1 & 2
5010 !  Get F(1,I), two-element DHTs
5020 FOR I = 0 TO N - 2 STEP 2
5030    F(1,I) = F(0,I) + F(0,I + 1)
5040    F(1,I + 1) = F(0,I) - F(0,I + 1)
5050 NEXT I
5060 IF P = 1 THEN L = 1 @ GOTO 170 !  Done
5070 !  Get F(2,I), four-element DHTs using table
5080 L = 2
5090 FOR I = 0 TO N - 4 STEP 4
5100    F(2,I) = F(1,I) + F(1,I + 2)
5110    F(2,I + 1) = F(1,I + 1) + F(1,I + 3)
5120    F(2,I + 2) = F(1,I) - F(1,I + 2)
5130    F(2,I + 3) = F(1,I + 1) - F(1,I + 3)
5140 NEXT I
5150 IF P = 2 THEN GOTO 170 !  Done
5160 RETURN
```

```
6000 !  Subr Stages 3,4...
6010 U = P7
6020 S = 4
6030 FOR L = 2 to P7
6040    S2 = S + S
6050    U = U - 1
6060    S0 = M(U - 1)
6070    FOR Q = 0 TO N7 STEP S2
6080       I = Q
6090       D = I + S
6100       F(L + 1,I) = F(L,I) + F(L,D)
6110       F(L + 1,D) = F(L,I) - F(L,D)
6120       K = D - 1
6130       FOR J = S0 TO N4 STEP S0
6140          I = I + 1
6150          D = I + S
6160          E = K + S
6170          X = F(L,E)*C(J)-F(L,D)*S(J)
6180          Y = F(L,D)*C(J) + F(L,E)*S(J)
6190          F(L + 1,I) = F(L,I) + Y
6200          F(L + 1,D) = F(L,I) - Y
6210          F(L + 1,K) = F(L,K) - X
6220          F(L + 1,E) = F(L,K) + X
6230          K = K - 1
6240       NEXT J
6250       E = K + S
6260    NEXT Q
6270    S = S2
6280 NEXT L
6290 RETURN

7000 !  Subr Get DFT
7010 R(0) = F(L,0) + F(L,0) @ X(0) = 0
7020 FOR I = 1 TO N2
7030    B = F(L,N - I)
7040    R(I) = F(L,I) + B
7050    X(I) = -F(L,I) + B
7060 NEXT I
7070 RETURN

8000 !  Subr Print results
8005 IF P > 1 THEN GOSUB 1000 !  Reread data
8010 CLEAR @ PRINT @ PRINT "τ,ν    f(τ)   H(ν)   R(ν) + jX(ν)"
      @ PRINT
8020 J$= ""
8030 FOR I = 0 TO N7
8040    J = MIN(I,N - I) ! Reflect
8050    F = INT(.5 + 1000*F(0,I))/1000 !  Round off
8060    H = INT(.5 + 1000/N*F(L,I))/1000
8070    R = INT(.5 + 1000*R(J)/N)/2000 @ X = INT(.5 + 1000*X(J)
         /N)/2000*SGN(N2 - I)
8080    J$= " +j " & VAL$ (ABS(X)) @   IF SGN(X) = -1 THEN
         J$ = " -j " & VAL$ (ABS(X))
8090    IF X = 0 THEN J$ = ""
```

```
8100    PRINT I; TAB(6); F; TAB(12); H; TAB(21); R; J$
8110 NEXT I
8120 PRINT @ PRINT " N = " & VAL$ (N) & ", Time was " & VAL$
      (TO) & " sec"
8130 RETURN
```

Sample printout with $n = 8$, $p = 3$

τ, ν	f(τ)	H(ν)	R(ν) + jX(ν)
0	1	4.5	4.5
1	2	-1.707	-.5 +j 1.207
2	3	-1	-.5 +j .5
3	4	-.707	-.5 +j .207
4	5	-.5	-.5
5	6	-.293	-.5 -j .207
6	7	0	-.5 -j .5
7	8	.707	-.5 -j 1.207

N=8. Time was 1.045 sec

Notes. (a) Substitute numerical value for $n = 2^p$ *in lines 20, 30 and 40 and value for* p *in line 50. (b) The variables used are* A B D E F H I J K L N N2 N4 N7 P P7 Q R S S0 S2 T T0 T6 T7 T8 T9 U W X Y C() F() M() R() S() X() J$. *(c)*

```
9000 !                    "FHTSUB"
9010 !  This subroutine takes input F() and returns the DHT in
      the same F()
9020 !  Declare DIM F(n), N = n and P = p before entering the
      subroutine
9030 IF P = 1 THEN J = F(0) + F(1) @    F(1) = F(0) - F(1) @
      F(0) = J @ RETURN
9040 DIM S9(n/4), T9(n/4), M9(10) !  sin θ, tan θ/2 & 2∧I
9050 DIM A9(64) !  Cell ordinate
9060 DIM V9(10), C9(10) !  Constants
9070 N9 = 2∧(P-2) @ N = 4*N9 @ C9(5) = N - 1 @ C9(6) = P - 1
9080 ON ERROR GOTO 9200 !  Forgives failure to initialize NO
      in main program
9090 IF N = NO THEN GOTO 9400 !  Skip pretabulation
9100 OFF ERROR ! Cancels future forgiveness

9200 !  Get powers of 2
9201 I = 1 @ M9(0) = 1 @ M9(1) = 2
9202 M9(I + 1) = M9(I) + M9(I) @ I = I + 1
9204 IF I < P THEN GOTO 9202

9296 IF N = 2 THEN GOTO 9411 !  Special case
9297 IF N < 8 THEN GOTO 9400 !  Skip trigonometric functions
```

```
9298 S9(N9) = 1
9299 IF N = 8 THEN S9(1) = SIN(PI/4) @ GOTO 9330 !  Skip sines

9300 !  Get sines
9301 FOR I = 1 TO 3 @   S9(I*N9/4) = SIN(I*PI/8) @   NEXT I !
      Coarse seed table for sines
9302 H9 = 1/2/COS(PI/16) !  Initial half secant
9303 !  Fill sine table
9304 C9(4) = P - 4
9305 FOR I = 1 TO P - 4
9306    C9(4) = C9(4) - 1 @ V9(0) = 0
9307    FOR J = M9(C9(4)) TO N9 - M9(C9(4)) STEP M9(C9(4) + 1)
9308       V9(1) = J + M9(C9(4))
9309       S9(J) = H9*(S9(V9(1)) + V9(0)) @ V9(0) = S9(V9(1))
9310    NEXT J
9311    H9 = 1/SQR(2 + 1/H9) !  Half secant recursion
9312 NEXT I

9330 !  Get tangents
9340 C9(0) = N9 - 1
9350 FOR I = 1 TO N9 - 1
9360    T9(I) = (1 - S9(C9(0)))/S9(I)
9370    C9(0) = C9(0) - 1
9380 NEXT I
9381 T9(N9) = 1

9400 !  Fast permute
9402 !  For P = 2, 3 permute directly
9403 IF P = 2 THEN V9(9) = F(1) @ F(1) = F(2) @ F(2) = V9(9)
      @ GOTO 9500
9404 IF P = 3 THEN V9(9) = F(1) @ F(1) = F(4) @ F(4) = V9(9)
      @ V9(9) = F(3) @ F(3) = F(6) @ F(6) = V9(9)
9405 IF P = 3 THEN GOTO 9500
9406 !  For P = 4, 5, 6 (Q9 = 2, 3), skip structure table
9407 Q9 = P DIV 2 @ C9(2) = M9(Q9)
9408 Q9 = Q9 + P MOD 2
9409 IF Q9 = 2 THEN A9(1) = 2 @   A9(2) = 1 @   A9(3) = 3 @
      GOTO 9420
9410 IF Q9 = 3 THEN A9(1) = 4 @   A9(2) = 2 @   A9(3) = 6 @
      A9(4) = 1 @ A9(5) = 5 @ A9(6) = 3 @ A9(7) = 7 @ GOTO 9420
9411 IF N = 2 THEN V9(6) = F(0) @ F(0) = F(1) @ F(1) = V9(6)
      !  Special case
9412 !  Set up structure table
9413 A9(0) = 0 @ A9(1) = 1
9414 FOR I = 2 TO Q9
9415    FOR J = 0 TO M9(I - 1) - 1
9416       A9(J) = A9(J) + A9(J)
9417       A9(J + M9(I - 1)) = A9(J) + 1
9418    NEXT J
9419 NEXT I

9420 !  Permute
9421 FOR I = 1 TO C9(2) - 1
9422    V9(4) = C9(2)*A9(I)
```

```
9423     V9(5) = 1 @ V9(6) = V9(4)
9424     V9(7) = F(V9(5)) @  F(V9(5)) = F(V9(6)) @  F(V9(6))
          = V9(7)
9425     FOR J = 1 TO A9(I) - 1
9426        V9(5) = V9(5) + C9(2) @ V9(6) = V9(4) + A9(J)
9427        V9(7) = F(V9(5)) @     F(V9(5)) = F(V9(6)) @
             F(V9(6)) = V9(7)
9428     NEXT J
9429 NEXT I

9500 !  Stages 1 & 2
9501 !  Get two-element DHTs
9502 FOR I = 0 TO N - 2 STEP 2
9503     V9(6) = F(I) + F(I + 1) @ V9(7) = F(I) - F(I + 1)
9505     F(I) = V9(6) @ F(I + 1) = V9(7)
9507 NEXT I
9508 IF P = 1 THEN RETURN ! Finished
9509 !  Get four-element DHTs
9510 FOR I = 0 TO N - 4 STEP 4
9511     V9(6) = F(I) + F(I + 2) @ V9(7) = F(I + 1) + F(I + 3)
9513     V9(8) = F(I) - F(I + 2) @ V9(0) = F(I + 1) - F(I + 3)
9515     F(I) = V9(6) @  F(I + 1) = V9(7) @  F(I + 2) = V9(8)
          @ F(I + 3) = V9(9)
9519 NEXT I
9520 IF P = 2 THEN RETURN ! Finished

9600 !  Stages 3, 4, ...
9601 U9 = C9(6) @ S9 = 4
9603 FOR L9 = 2 TO C9(6)
9604     V9(2) = S9 + S9 @ U9 = U9 - 1 @ V9(3) = M9(U9 - 1)
9607     FOR Q9 = 0 TO C9(5) STEP V9(2)
9608        I = Q9 @ D9 = I + S9
9610        V9(6) = F(I) + F(D9) @ V9(7) = F(I) - F(D9)
9612        F(I) = V9(6) @ F(D9) = V9(7) @ K9 = D9 - 1
9615        FOR J = V9(3) TO N9 STEP V9(3)
9616           I = I +1 @ D9 = I + S9 @ E9 = K9 + S9
9617           V9(9) = F(D9) + F(E9)*T9(J)
9618           X9 = F(E9) - V9(9)
9621           V9(6) = F(I) + Y9 @    V9(7) = F(I) - Y9 @
                 V9(8) = F(K9) - X9 @ V9(9) = F(K9) + X9
9625           F(I) = V9(6) @  F(D9) = V9(7) @  F(K9) = V9(8)
                 @ F(E9) = V9(9) @ K9 = K9 - 1
9630        NEXT J
9631        E9 = K9 + S9
9632     NEXT Q9
9633     S9 = V9(2)
9634 NEXT L9
9635 N0 = N ! Remember
9636 RETURN
```

Notes: (a) Substitute numerical values for n in line 9040. (b) Declare **DIM F(*n*)**, **N** = *n and* **P** = *p before entering the subroutine. (c) This subroutine assumes that user will divide throughout by N subsequently. (d) With the exception of* **F() I, J, K, N** *and* **P**, *variable names in this subroutine are mainly the same as those used in* **FHTBAS** *with a 9 appended, which makes it*

easy to avoid duplication. The list of variables is D9 E9 H9 I J K L9 N NO P Q9 S9 U9 X9 Y9 C9() F() M9() S9() T9() V9().
(e) Line 9635 remembers N. *User may wish to start by mentioning in the main program that* NO = 0 *so that line 9190 will not stop execution; then lines 9080 and 9100, which take care of failure to mention* NO, *may be deleted.*

```
10 !                          "FHTPS"
20 !  Computes the power spectrum P() of a data sequence
      F() via the DHT
30 N = 64 @ P = 6 !  Needed for Subr 9000
40 DIM F(64), P(32)
50 D = 1 !  Adjust for degree of smoothing

100 !  Main program
110 GOSUB 1000 !  Insert data
120 GOSUB 9000 !  Get DHT
130 GOSUB 2000 !  Get power spectrum
140 GOSUB 3000 !  Smooth the power spectrum
150 GOSUB 4000 !  Print the results
160 END

1000 !  Subr Insert data
1010 !  Bandpass noise generator
1020 F(0) = RND @ F(1) = RND
1030 FOR I = 2 TO N
1040    F(I) = 1.65*F(I - 1) - 0.95*F(I - 2) + RND - .5
1050 NEXT I
1060 RETURN

2000 !  Subr Get power spectrum
2010 P(0) = 2*F(0)^2
2020 FOR I = 1 TO N/2 - 1
2030    P(I) = F(I)^2 + F(N - I)^2
2040 NEXT I
2050 RETURN

3000 !  Subr Smooth the power spectrum
3010 IF D = 0 THEN RETURN
3020 FOR I = 1 TO D
3030    K = N/2 - 1 - I
3040    FOR J = 0 TO K @ P(J) = P(J) + P(J + 1) @ NEXT J
3050    FOR J = 0 TO K - 1 @        P(K - J) = P(K - J) +
         P(K - J - 1) @ NEXT J
3060    P(0) = 2*P(0)^2
3070 NEXT I
3080 RETURN

4000 !  Subr Print the results
4010 PRINT "ν        P(ν)" @ PRINT
```

```
4020 FOR I = 0 TO N/2 - 1 - D
4030     PRINT I; P(I)/4∧D/N/N/2
4040 NEXT I
4050 RETURN

9000 !  Subr Get DHT (Listed elsewhere)
```

Sample printout for $D = 1$

ν	$P(\nu)$
0	.007
1	.007
2	.004
3	.025
4	.062
5	.157
6	.238
7	.140
8	.026
9	.007
10	.003
11	.003
12	.003
13	.003
14	.003
15	.002
...	...
28	.000
29	.000
30	.001

Notes. (a) Declare N *and* P *as desired in line 30. (b) Subroutine 1000 generates a data sequence representing bandpass noise. (c) Subr 2000 gets the raw power spectrum. (d) Subr 3000 smoothes the power spectrum by a fast, elegant binomial smoothing algorithm. Adjust degree of smoothing* D *as desired on line 50. With* D = 1, *the coefficients are* {1 2 1 }/4; D = 2 *gives* {1 4 6 4 1}/16 *and so on. The factors 4, 16, etc. are allowed for by the factor* 4∧D *in the printout subroutine, which also allows for the factor of 2 suppressed in line 2020 and the factor* N *suppressed in subroutine 9000.*

```
10 !                     "CONV"
20 !  Performs convolution of F1() with F2() to yield con-
      volution sum C()
30 !  Sequences F1() and F2() have lengths l1 and l2 to be
      declared.  C() will have length l1 + l2 - 1

100 !  State dimensions
110 !  DIM F1(l1), F2(l2), C(l1 + l2 - 1)
120 !  Declare lengths
130 L1 = l1 !  Length of F1()
140 L2 = l2 !  Length of F2()
```

```
150 L3 = L1 + L2 - 1 !  Length of C()
160 GOSUB 1000 !  Insert data
1700 GOSUB 2000 !  Convolve
180 GOSUB 3000 !  Print results
190 END

500 DATA 1, 2, 3, 4, 5
510 DATA 1, 2, 2, 4

1000 !  Subr Insert data
1010 FOR I = 1 TO L1 @ READ F1(I) @ NEXT I
1020 FOR I = 1 TO L2 @ READ F2(I) @ NEXT I
1030 RETURN

2000 Subr Convolve
2010 FOR I = 1 TO L3
2020    C(I) = 0 !  Prepare to sum
2030    FOR J = MAX(1 - L2 + I, 1) TO MIN(L1, I)
2040       K = I - J + 1
2050       C(I) = C(I) + F1(J)*F2(K)
2060    NEXT J
2070 NEXT I
2080 RETURN

3000 !  Subr Print results
3010 PRINT "F1(I)  F2(I)  C(I)"
3020 PRINT
3030 FOR I = 1 TO L3
3040   F1$ = " " @ IF I <= L1 THEN F1$ = VAL$ (F1(I))
3050   F2$ = " " @ IF I <= L2 THEN F2$ = VAL$ (F2(I))
3060   PRINT F1$ ; TAB(8); F2 $; TAB(14); C(I)
3070 NEXT I
3080 RETURN
```

Sample printout with $l_1 = 5$ and $l_2 = 2$

```
F1(I)  F2(I)  C(I)

1      1      1
2      2      4
3      2      9
4      4      18
5             27
.             30
.             26
.             20
```

Notes. *(a) Substitute numerical values for l_1 and l_2 in lines 110, 130 and 140. For the sample printout, $l_1 = 5$ and $l_2 = 4$ were used. (b) Data may be inserted by formula rather than through* **DATA** *statements, e.g.*

```
1010 FOR I = 1 TO L1 @ F1(I) = I @ NEXT I
```

or by assignment statements, e.g.

```
1020 F2(1) = 1 @ F2(2) = 2 @ F2(3) = 2 @ F2(4) = 4
```

(c) For $f_1 \star f_2$, the cross-correlation of f_1 on f_2, rewrite line 1010 as

```
1010 FOR I = L1 TO 1 STEP -1 @ F1(I) = I @ NEXT I
```

```
10 !                          "CCONV"
20 !  Performs cyclic convolution of F1() with F2() to yield
      C()
30 !  Sequences F1() and F2() may have any lengths l1 and l2
40 !  Declare N to be equal to or greater than the larger of
      l1 and l2
50 N = n
60 DIM F1(n), F2(n), C(n)
70 !  Initialize
80 FOR I = 0 TO N - 1 @ F1(I), F2(I), C(I) = 0 @ NEXT I

100 !  Main program
110 GOSUB 1000 !  Insert data
120 GOSUB 2000 !  Convolve
130 GOSUB 3000 !  Print results
140 END

1000 !  Subr Insert data
1010 FOR I = 0 TO N - 1
1020    F1(I) = 1/2^ I
1030 NEXT I

1040 F2(0) = 256
1050 F2(1) = -128
1060 F2(2) = -128
1070 RETURN

2000 !  Subr Convolve
2010 FOR S = 0 TO N - 1 !  Each shift in turn as F1() moves
2020    C(S) = 0 !  Prepare to sum
2030    FOR T = 0 TO N - 1
2040       U = (S - T) MOD N ! Moving index is cycled
2050       C(S) = C(S) + F1(U)*F2(T) ! Sum products
2060    NEXT T
2070 NEXT S
2080 RETURN

3000 !  Subr Print results
3010 PRINT "I F1(I) F2(I) C(I)"
3020 PRINT
3030 FOR I = 0 TO N - 1
3040    PRINT I; TAB(6); F1(I); TAB(16); F2(I); C(I)
3050 NEXT I
3060 RETURN
```

Sample printout with $n = 8$

```
I    F1(I)     F2(I)   C(I)

0    1          256     253
1    .5        -128    -1
2    .25       -128    -128
3    .125       0      -64
4    .0625      0      -32
5    .03125     0      -16
6    .015625    0      -8
7    .0078125   0      -4
```

Notes. (a) For the data illustrated, $n = 8$. The desired numerical value of n must be substituted in lines 50 and 60. (b) Data may be inserted by DATA *statements as in* CONV *if appropriate; Subr 1000 illustrates alternative methods. (c) For the cyclic cross-correlation $f_1 \circledast f_2$, reverse f_1.*

```
10 !                          "MATCON"
20 !  Performs convolution of A() with B() by matrix multi-
        plication to yield convolution sum C()
30 L1 = l1 @ L2 = l2
40 DIM A(l1 + l2 - 2, l2 - 1), B(l2 - 1), C(l1 + l2 - 2)
50 L3 = L1 + L2 - 1
60 !  Read column matrix [B]
70 FOR I = 0 TO L2 - 1
80    READ B(I)
90 NEXT I
100 !  Fill rectangular matrix [A]
110 FOR I = 0 TO L3 - 1
120    IF I < L1 THEN READ A(I,0) ELSE A(I,0) = 0 !  Col. 1
130    FOR J = 1 TO L2 - 1
140       A((I + J) MOD L3,J) = A(I,0)
150    NEXT J
160 NEXT I
170 !  Multiply [A]x[B]
180 MAT C = A*B
190 !  Print results
200 FOR I = 0 TO L3 - 1
210    PRINT C(I)
220 NEXT I
230 END
240 !  Column matrix [B]
250 DATA 1,2,3
260 !  Column 1 of matrix [A]
270 DATA 1,1,1,1,2
280 END
```

Notes. (a) Substitute numerical values for l_1, the length of sequence A()*, and for l_2, the length of sequence* B()*, in lines 30 and 40. The sample used $l_1 = 5$ and $l_2 = 3$. (b) The circulant matrix $[A]$ corresponding to sequence* A()*, and the column matrices in the equation $[C] = [A] \times [B]$ are shown here in full:*

$$\begin{bmatrix}1\\3\\6\\6\\7\\7\\6\end{bmatrix} = \begin{bmatrix}1&0&0\\1&1&0\\1&1&1\\1&1&1\\2&1&1\\0&2&1\\0&0&2\end{bmatrix} \times \begin{bmatrix}1\\2\\3\end{bmatrix}$$

```
10 !                          "ACF"
20 !  Performs simple autocorrelation of F() to yield C()
30 !  Sequence F() has length l1 and C() has length 2l1 - 1
40 L1 = l
50 L3 = 2*L1 - 1 !  Length of C()
60 !  DIM F(l1), C(2l1 - 1)

100 GOSUB 1000 !  Insert data
110 GOSUB 2000 !  Autocorrelate
120 GOSUB 3000 !  Print results
130 END

500 DATA 1, 4, 6, 4, 1

1000 !  Subr Insert data
1010 FOR I = 1 TO L1 @ READ F(I) @ NEXT I
1020 RETURN

2000 !  Subr Autocorrelate
2010 FOR I = 1 TO L3
2020    C(I) = 0 !  Prepare to sum
2030    FOR J = MAX(1 - L1 +I, 1) TO MIN(L1, I)
2040       K = J - I + L1
2050       C(I) = C(I) + F(J)*F(K)
2060    NEXT J
2070 NEXT I
2080 RETURN

3000 !  Subr Print results
3010 PRINT " I     F(I)     C(I)"
3020 PRINT
3030 FOR I = 1 TO L3
3040    F$ "" @ IF I <= L1 THEN F$ = VAL$(F(I))
3050    PRINT I; TAB(7); F$; TAB(12); C(I)
3060 NEXT I
3070 RETURN
```

Sample printout with $l_1 = 5$

I	F(I)	C(I)
1	1	1
2	4	8
3	6	28
4	4	56
5	1	70
6		56
7		28
8		8
9		1

Notes. (a) *Substitute numerical value for* l_1 *in lines 40 and 60. Example uses* $l_1 = 5$.

```
10 !                          "CACF"
20 !  Performs cyclic autocorrelation of F() to yield C()
30 N = n
40 DIM F(n), C(n)

100 GOSUB 1000 !  Insert data
110 GOSUB 2000 !  Autocorrelate
120 GOSUB 3000 !  Print results
130 END

500 DATA 1, 1, 1, 1, 0, 0, 0, 0

1000 !  Subr Insert data
1010 FOR I = 0 TO N - 1 @ READ F(I) @ NEXT I
1020 RETURN

2000 !  Autocorrelate
2010 FOR S = 0 TO N - 1 !  Each shift in turn
2020    C(S) = 0 !  Prepare to sum
2030    FOR T = 0 TO N - 1
2040       U = (T - S) MOD N ! Shifted index is cycled
2050       C(S) = C(S) + F(U)*F(T) ! Sum products
2060    NEXT T
2070 NEXT S
2080 RETURN

3000 !  Subr Print results
3010 PRINT " I     F(I)    C(I)" @ PRINT
3020 FOR I = 0 TO N - 1
3030    PRINT I; TAB(6); F(I); TAB(12); C(I)
3040 NEXT I
3050 RETURN
```

Sample printout with $n = 8$

```
I    F(I)    C(I)

0    1       4
1    1       3
2    1       2
3    1       1
4    0       0
5    0       1
6    0       2
7    0       3
```

Notes. Substitute numerical value for n in lines 30 and 40. Example uses $n = 8$.

```
10 !                          "ICONV"
20 !  Inverts convolution F1()*F2() = G() to find F1() when
      F2() and G() are given
30 L1 = l1 !  Number of terms of F1() to be found
```

```
40 L2 = l2 !  Length of F2()
50 L3 = l3 !  Length of G()
60 DIM F1(l1 - 1), F2(l2 - 1), G(l1 + l2 - 2)

100 GOSUB 1000 !  Insert data
110 GOSUB 2000 !  Deconvolve
120 GOSUB 3000 !  Print results
130 END

500 DATA 2,2,3,3,4
510 DATA 2,4,9,10,13,10,8

1000 !  Subr Insert data
1010 FOR I = 1 TO L2 @ READ F2(I) @ NEXT I
1020 FOR I = 1 TO L3 @ READ G(I) @ NEXT I
1030 FOR I = L3 TO L1 + L2 - 2 @ G(I) = 0 @ NEXT I
1040 RETURN

2000 !  Subr Deconvolve
2010 F1(1) = G(1)/F2(1)
2020 FOR I = 2 TO L1
2030    S = 0 !  Prepare to sum
2040    FOR J = MAX(1,I + 1 - L2) TO I - 1
2050       S = S + F1(J)*F2(I + 1 - J)
2060    NEXT J
2070    F1(I) = (G(I) - S)/F2(0)
2080 NEXT I
2090 RETURN

3000 !  Subr Print results
3010 PRINT " I    F1(I)    F2(I)    G(I)" @ PRINT
3020 FOR I = 1 TO L1
3030    F2$ = " " @ IF I <= L2 - 1 THEN F2$ = VAL$ (F2(I))
3040    G$ = " " @ IF I <= L3 - 1 THEN G$ = VAL$ (G(I))
3050    PRINT I; TAB(6); F1(I); TAB(14); F2$ ; TAB(22); G$
3060 NEXT I
3070 RETURN
```

Sample printout with $l_1 = 8$, $l_2 = 5$ *and* $l_3 = 7$

```
I    F1(I)  F2(I)   G(I)

1    1      2       2
2    1      2       4
3    2      3       9
4    0      3       10
5    0      4       13
6    0              10
7    0              8
8    0
```

Notes. (a) Substitute values for l_1, l_2, l_3 *in lines 30, 40, 50, 60. The example uses* $l_1 = 8$, $l_2 = 5$, $l_3 = 7$. *(b) This program implements a direct hand-calculation algorithm that was originally presented on p. 36 of R.N. Bracewell, The Fourier Transform and its Applications (McGraw-Hill, 1965).*

```
10 !                        "RECIP"
20 !  Obtains reciprocal sequence I() for given sequence F()
30 L1 = l1 !  Number of terms of I() to be found
40 L2 = l2 !  Length of F()
50 DIM I(l1 - 1), F(l2 - 1)

100 GOSUB 1000 !  Insert data
110 GOSUB 2000 !  Invert
120 GOSUB 3000 !  Print results
130 END

500 DATA 1,1,1,1,1

1000 !  Subr Insert data
1010 FOR I = 0 TO L2 - 1 @ READ F(I) @ NEXT I
1020 RETURN

2000 !  Subr Invert
2010 I(0) = 1/F(0)
2020 FOR I = 1 TO L1 - 1
2030    S = 0 !  Prepare to sum
2040    FOR J = MAX(0,I + 1 - L2) TO I - 1
2050       S = S + I(J)*F(I - J)
2060    NEXT J
2070    I(I) = - S/F(0)
2080 NEXT I
2090 RETURN

3000 !  Subr Print results
3010 PRINT " I F(I) I(I)" @ PRINT
3020 FOR I = 0 TO L1 - 1
3030    F$ = " " @ IF I < = L2 - 1 THEN F$ = VAL$ (F(I))
3040    PRINT I; TAB(7); F$ ; TAB(14); I(I)
3050 NEXT I
3060 RETURN
```

Sample printout with $l_1 = 8$ and $l_2 = 5$

```
I   F(I)   I(I)

1    1       1
2    1      -1
3    1       0
4    1       0
5    1       0
6            1
7           -1
8            0
```

Notes. (a) Substitute values for l_1 and l_2 in lines 30, 40, and 50. The example uses $l_1 = 8$, $l_2 = 5$. (b) This algorithm is a special case of **ICONV**, *where $G(I) = \{1\ 0\ 0\ ...\}$.*

```
10 !                          "ICORR"
20 DIM A(2h+1 ), F(2h+1 )
30 H = h !  Number of data items
40 N = 2*H - 1 !  Full length of autocorrelation

100 GOSUB 1000 !  Insert data
110 GOSUB 4000 !  Invert correlation
120 GOSUB 5000 !  Print results
130 END

500 DATA 20, 15, 6, 1 !  Binomial coefficients

1000 !  Subr Insert data
1010 FOR I = 0 TO H - 1 @ READ A(I + H - 1) @ A(H - 1 - I)
    = A(I + H - 1)@ NEXT I
1020 A0 = A(0)
1030 FOR I = 0 TO N - 1 @ A(I) = A(I)/A0 @ NEXT I
1040 RETURN

4000 !  Subr Invert correlation
4010 F(0) = 1
4020 F(1) = A(1)/2
4030 F(2) = A(2)/2 - F(1)*F(1)/2
4040 FOR I = 3 TO N - 1 !  Subscript
4050    K = I DIV 2 !  Number of steps
4060    S = 0 !  Prepare to sum
4070    FOR J = 1 TO K - 1
4080       S = S + F(J)*F(I - J)
4090    NEXT J
4100    IF I MOD 2 = 0 THEN S = S + F(K)*F(K)/2
        ELSE S = S + F(K)*F(I - K) ! Last term
4110    F(I) = A(I)/2 - S
4120 NEXT I
4130 RETURN

5000 !  Subr Print results
5010 PRINT "I A() F()" @ PRINT
5020 FOR I = 0 TO N - 1
5030    IF I > N - 1 THEN A$ = " " ELSE A$ = VAL$(A(I))
5040    PRINT I; TAB(6); A$; TAB(12); F(I)
5050 NEXT I
5060 RETURN
```

Sample printout

```
I    A()  F()

0     1    1
1     6    3
2    15    3
3    20    1
4    15    0
5     6    0
6     1    0
```

Notes. *(a) Substitute numerical value for* h *in line 30, and in line 20 insert value of* $2h+1$. *(b) The algorithm implements the analogue calculator method where* $A(\,)$ *would be written on the outer rim and* $F(\,)$ *would appear step by step on both the inner rings, but in opposite directions. At each step, as values of* $F(\,)$ *become known, they are incorporated. (c) Unless the given* $A(\,)$ *is the autocorrelation of a finite sequence, as in the example, the series* $F(\,)$ *will continue indefinitely. (d) The word causal is used in the sense* $A(i) = 0$, $i < 0$, *a significant condition in physics but automatically implicit in a computer. (e) It is generally understood that autocorrelation functions cannot be unambiguously inverted. The present algorithm gives the even solution of finite support, if there is one. However, it begins on the left hand term of* $A(\,)$ *which may be a small and error-afflicted member of a data sequence. Change* $A(0)$ *slightly, say to 1.1, and renormalize, to observe the error propagation.*

```
10 !                          "FHTCONV"
20 !  Computes the convolution of two functions of length
      N/2 using the fast algorithm
30 P = 4 @ N = 2^P
40 DIM F(16), T(16)

100 GOSUB 1000 !  Get first function
110 GOSUB 9000 !  Get DHT
120 GOSUB 2000 !  Shift to temporary array
130 GOSUB 3000 !  Get second function
140 GOSUB 9000 !  Get DHT
150 GOSUB 4000 !  Multiply DHTs
160 GOSUB 9000 !  Get DHT
170 GOSUB 5000 !  Print results
180 END

1000 !  Subr Get first function
1010 FOR I = 0 TO N - 1
1020 IF I < 4 OR I > 12 THEN F(I) = 1 ELSE F(I) = 0
1030 NEXT I
1040 RETURN

2000 !  Subr Shift to temporary array
2010 FOR I = 0 TO N - 1
2020     T(I) = F(I)
2030 NEXT I

3000 !  Subr Get second function
3010 FOR I = 0 TO N - 1
3020     IF I < 8 THEN F(I) = I + 1 ELSE F(I) = 0
3030 NEXT I
3040 RETURN
```

```
4000 !  Subr Multiply DHTs
4010 FOR I = 0 TO N - 1
4020    F(I) = F(I)*T(I)
4030 NEXT I
4040 RETURN

4400 !  Subr Symmetry absent
4410 DIM F1(16) !  Reverse of F()
4420 FOR Z = 1 TO N - 1
4430    F1(Z) = F(N - Z)
4440 NEXT Z ! Store reverse of F()
4450 F(0) = F(0)*T(0)*2
4460 FOR I = 1 TO N - 1
4470    F(I) = F(I)*(T(I) + T(N - I)) + F1(I)*(T(I)
        - T(N - I))
4480 NEXT I
4490 RETURN

5000 !  Subr Print results
5010 FOR I = 0 TO N - 1
5020    PRINT INT(0.5 + F(I)/N)
5030 NEXT I
5040 RETURN

9000 !  Subr Get DHT (Listed elsewhere)
```

Sample printout

```
10 15 21 28 35 33 30 26 21 15 8 0 0 1 3 6
```

Notes. (a) Declare the functions in subroutines 1000 and 3000 and pack each to length N *with zeros. (b) Declare numerical value of* P *in line 30. (c) Substitute numerical value* N *for n in lines 30 and 9040. (d) Replace 4000 by 4400 in line 150 if neither of the functions has symmetry; and in line 5020 replace* F(I)/N *by* F(I)/N/2.

```
10 !                        "FHTACF"
20 !  Computes the autocorrelation of a function of length
      N/2 using the fast algorithm
30 P = p @ N = 2^P
40 DIM F(n), P(n)

100 GOSUB 1000 !  Declare the function
110 GOSUB 9000 !  Get DHT
120 GOSUB 2000 !  Get power spectrum
130 GOSUB 9000 !  Get DHT
140 GOSUB 5000 !  Print results
150 END

1000 !  Subr Declare the function
1010 FOR I = 0 TO N - 1
```

```
1020    IF I < 7 THEN F(I) = 1 ELSE F(I) = 0
1030 NEXT I
1040 RETURN

2000 !  Subr Get power spectrum
2010 P(0) = 2*F(0)∧2
2020 FOR I = 1 TO N/2
2030    P(I), P(N - I) = (F(I)∧2 + F(N - I)∧2
2040 NEXT I
2050 FOR I = 0 TO N - 1 @ F(I) = P(I) @ NEXT I
2060 RETURN

5000 !  Subr Print results
5010 FOR I = 0 TO N - 1
5020    PRINT F(I)/N/2
5030 NEXT I
5040 RETURN

9000 !  Subr Get DHT (Listed elsewhere)
```

Sample printout for a string of seven 1s and $n = 16$ $(p = 4)$

```
7 6 5 4 3 2 1 0 0 0 1 2 3 4 5 6
```

Notes. (a) Declare the N/2 *values of the function in subroutine 1000 and pack with zeros to length* N. *(b) Declare numerical value of* p *in line 40. (c) Substitute numerical value* N *for* n *in lines 30 and 9040.*

```
10 !                      "FHTRX4"
20 !  Radix-4 calculation of FHT of F()
30 P = p
40 N = 4∧P @ N4 = N/4
50 R = SQR(2)
60 DIM F(n)

100 GOSUB 1000 !  Insert data
110 GOSUB 4000 !  Permute to radix 4
120 GOSUB 5000 !  Get DHT
130 GOSUB 8000 !  Print results
140 END

1000 !  Subr Insert data
1010 FOR I = 0 TO N - 1 @ F(I) = I + 1 @ NEXT I
1020 RETURN

4000 !  Subr Permute to radix 4
4010 J = 1 @ I = 0
4020 I = I + 1 @ IF I >= J THEN GOTO 4040
4030 T = F(J - 1) @ F(J - 1) = F(I - 1) @ F(I - 1) = T
4040 K = N4
4050 IF 3*K >= J THEN GOTO 4070
```

```
4060 J = J - 3*K @ K = K/4 @ GOTO 4050
4070 J = J + K @ IF I < N - 1 THEN GOTO 4020
4080 RETURN

5000 !  Subr Get DHT
5010 !  Stage 1
5020 FOR I = 0 TO N - 1 STEP 4
5030    T1 = F(I) + F(I + 1)
5040    T2 = F(I) - F(I + 1)
5050    T3 = F(I + 2) + F(I + 3)
5060    T4 = F(I + 2) - F(I + 3)
5070    F(I) = T1 + T3
5080    F(I + 1) = T1 - T3
5090    F(I + 2) = T2 + T4
5100    F(I + 3) = T2 - T4
5110 NEXT I

5130 !  Stages 2 to P
5140 FOR L = 2 TO P
5150    D1 = 2∧(L + L - 3)
5160    D2 = D1 + D1
5170    D3 = D2 + D2
5180    D4 = D2 + D3
5190    D5 = D3 + D3
5200    FOR J = 0 TO N - 1 STEP D5
5210       T1 = F(J) + F(J + D2)
5220       T2 = F(J) - F(J + D2)
5230       T3 = F(J + D3) + F(J + D4)
5240       T4 = F(J + D3) - F(J + D4)
5250       F(J) = T1 + T3
5260       F(J + D2) = T1 - T3
5270       F(J + D3) = T2 + T4
5280       F(J + D4) = T2 - T4
5290       T1 = F(J + D1)
5300       T2 = F(J + D1 + D2)*R
5310       T3 = F(J + D1 + D3)
5320       T4 = F(J + D1 + D4)*R
5330       F(J + D1) = T1 + T2 + T3
5340       F(J + D1 + D2) = T1 - T3 + T4
5350       F(J + D1 + D3) = T1 - T2 + T3
5360       F(J + D1 + D4) = T1 - T3 - T4
5370       FOR K = 1 TO D1 - 1
5380          L1 = J + K
5390          L2 = L1 + D2
5400          L3 = L1 + D3
5410          L4 = L1 + D4
5420          L5 = J + D2 - K
5430          L6 = L5 + D2
5440          L7 = L5 + D3
5450          L8 = L5 + D4
5460          A1 = PI*K/D3 @ A2 = A1 + A1 @ A3 = A1 + A2
5470          C1 = COS(A1) @ C2 = COS(A2) @ C3 = COS(A3)
               @ S1 = SIN(A1) @ S2 = SIN(A2) @ S3 = SIN(A3)
5480          T5 = F(L2)*C1 + F(L6)*S1
```

```
5490            T6 = F(L3)*C2 + F(L7)*S2
5500            T7 = F(L4)*C3 + F(L8)*S3
5510            T8 = F(L6)*C1 - F(L2)*S1
5520            T9 = F(L7)*C2 - F(L3)*S2
5530            T0 = F(L8)*C3 - F(L4)*S3
5540            T1 = F(L5) - T9
5550            T2 = F(L5) + T9
5560            T3 = - T8 - T0
5570            T4 = T5 - T7
5580            F(L5) = T1 + T4
5590            F(L6) = T2 + T3
5600            F(L7) = T1 - T4
5610            F(L8) = T2 - T3
5620            T1 = F(L1) + T6
5630            T2 = F(L1) - T6
5640            T3 = T8 - T0
5650            T4 = T5 + T7
5660            F(L1) = T1 + T4
5670            F(L2) = T2 + T3
5680            F(L3) = T1 - T4
5690            F(L4) = T2 - T3
5700        NEXT K
5710    NEXT J
5720 NEXT L
5730 RETURN

8000 !  Subr Print results
8010 PRINT " ν    H(ν)" @ PRINT
8020 FOR I = 0 TO N - 1 @ PRINT I; TAB(6); F(I)/N @ NEXT I
8030 RETURN
```

Notes. (a) Subtitute numerical values for p *in line 30 and* n *in line 60, such that length* $n = 4^p = 16, 64, 256, 1024, ...$ *(b) See Chapter 8 for discussion. (c) Pretabulation of trigonometric functions as in* FHTSUB *will speed computation. (d) Copyright © 1985 The Board of Trustees of the Leland Stanford Junior University.*

```
10 !                      "FASTPERMUTE"
20 !  Demonstrates fast permutation
30 !  NO is the number of cells per side and also the grid
      spacing within each cell
40 DIM F(1024), J(16)
50 P = p @ N8 = 2∧(P - 3)
60 N4 = N8 + N8 @ N2 = N4 + N4 @ N = N2 + N2
70 P1 = P @ N1 = N ! Family I
80 IF P MOD 2 = 1 THEN N1 = N2 @ P1 = P - 1 !  Family II
90 NO = SQR(N1/4) !  Cells per side
100 S = N/NO !  Cell size

150 GOSUB 1000 !  Subr Insert data
160 GOSUB 4000 !  Subr Get permutation function J(I)
```

```
170 GOSUB 5000 !  Subr Get cell coordinates X,Y and permute
180 GOSUB 9000 !  Subr Print results
190 END

1000 !  Subr Insert data
1010 FOR I = 0 TO N - 1 @ F(I) = I @ NEXT I
1020 RETURN

4000 !  Subr Get J(I)
4010 P2 = P1/2 - 1 @ J = 0
4020 J(0) = 0 @ J(N0 - 1) = N0 - 1
4030 FOR I0 = 1 TO N0 - 2 @ P0 = P2 - 1
4040    IF J < 2∧P0 THEN GOTO 4060
4050    J = J - 2∧P0 @ P0 = P0 - 1 @ GOTO 4040
4060    J = J + 2∧P0 @ J(I0) = J
4070 NEXT I0
4080 RETURN

5000 !  Subr Get cell coordinates (X,Y)
5010 S9 = -S
5020 FOR J = 0 TO N0 - 1
5030    S9 = S9 + S
5040    X = J(J) - S + J*S
5050    FOR I = J TO N0 - 1
5060       X = X + S
5070       Y = J(I) + S9
5080       GOSUB 8000 !  Permute cell contents
5090    NEXT I
5100 NEXT J
5110 RETURN

8000 !  Subr Permute cell contents
8010 !  Diagonal cells first
8020 IF X = Y AND P MOD 2 = 0 THEN T = F(X + N0) @F(X + N0)
       = F(Y + 2*N0) @F(Y + 2*N0) = T @ RETURN ! Family I

8030 IF X = Y AND P MOD 2 = 1 THEN T = F(X + N0) @F(X + N0)
       = F(Y + 4*N0) @ F(Y + 4*N0) = T
8040 IF X = Y AND P MOD 2 = 1 THEN T = F(X + 3*N0) @ F(X +
       3*N0) = F(Y + 6*N0) @   F(Y + 6*N0) = T @   RETURN
       !  Family II

8050 !  Off-diagonal cells
8060 A = X - N0
8070 FOR I9 = 0 TO 3 + 4*(P MOD 2)
8080    IF P MOD 2 = 1 THEN I2 = I9 + 3*((I9=1) + (I9=3)
       - (I9=4) - (I9=6)) ELSE I2 = I9 + (I9=1) - (I9=2)
8090    A = A + N0
8100    B = Y + I2*N0
8110    T = F(A) @ F(A) = F(B) @ F(B) = T ! Swap
8120 NEXT I9
8130 RETURN

9000 !  Subr Print results
9010 PRINT " i j" @ PRINT
```

```
9020 FOR I = 0 TO N - 1
9030    PRINT I; F(I)
9040 NEXT I
9050 RETURN
```

Sample printout with $p = 3$

```
i j

0 0
1 4
2 2
3 6
4 1
5 5
6 3
7 7
```

Notes: (a) In line 80 state the value of P *and in line 70 dimension* F() *to the corresponding value* 2^p. *(b)* DIM J(16) *takes care of* N < *4096; for larger* N *insert* NO *as dimension of* J(). *(c) See Chapter 8 for discussion.*

```
C  FHTBAS.FOR
C
      DIMENSION F(Ø:8,Ø:256), R(Ø:256), X(Ø:256), M(Ø:2Ø)
      DIMENSION S(64), C(64)
      INTEGER D, E, I, J, K, L, MØ, N, N2, N4
      INTEGER N7, P, PØ, P7, Q, S1, SØ, S2, S4, U
      FIN(I)=I+1.
      P=8
      N4=2**(P-2)
      N2=N4+N4
      N=N2+N2
      N7=N-1
      P7=P-1

C  INSERT DATA
      DO I=Ø,N7
      F(Ø,I)=FIN(I)
      F(1,I)=F(Ø,I)
      END DO

C  GET POWERS OF 2
      I=1
      M(Ø)=1
      M(1)=2
1Ø    M(I+1)=M(I)+M(I)
      I=I+1
      IF (I.LT.P) GO TO 1Ø
```

```
C  GET SINES AND COSINES
      PI=3.14159265
      W=2*PI/N
      A=Ø
      DO I=1,N4
      A=A+W
      S(I)=SIN(A)
      C(I)=COS(A)
      END DO

C  PERMUTE
      J=-1
      I=-1
2Ø    I=I+1
      PØ=P
3Ø    PØ=PØ-1
      J=J-M(PØ)
      IF (J.GE.-1) GO TO 3Ø
      J=J+M(PØ+1)
      IF (I.LE.J) GO TO 2Ø
      T=F(Ø,I+1)
      F(Ø,I+1)=F(Ø,J+1)
      F(Ø,J+1)=T
      IF (I.LT.(N-3)) GO TO 2Ø

C  GET F(I,1), 2-ELEMENT DHTs
      DO I=Ø,N-2,2
      F(1,I)=F(Ø,I)+F(Ø,I+1)
      F(1,I+1)=F(Ø,I)-F(Ø,I+1)
      END DO
      IF (P.EQ.1) GO TO 4Ø

C  GET F(2,I), 4-ELEMENT DHTs
      L=2
      MØ=2
      DO I=Ø,N-4,4
      F(2,I)=F(1,I)+F(1,I+2)
      F(2,I+1)=F(1,I+1)+F(1,I+3)
      F(2,I+2)=F(1,I)-F(1,I+2)
      F(2,I+3)=F(1,I+1)-F(1,I+3)
      END DO
      IF (P.EQ.2) GO TO 4Ø

C  STAGES 3,4,...
      U=P7
      S1=4
      DO L=2,P7
      S2=S1+S1
      U=U-1
      SØ=M(U-1)
      DO Q=Ø,N7,S2
      I=Q
      D=I+S1
      F(L+1,I)=F(L,I)+F(L,D)
```

```
      F(L+1,D)=F(L,I)-F(L,D)
      K=D-1
      DO J=SØ,N4,SØ
      I=I+1
      D=I+S1
      E=K+S1
      Y=F(L,D)*C(J)+F(L,E)*S(J)
      Z=F(L,D)*S(J)-F(L,E)*C(J)
      F(L+1,I)=F(L,I)+Y
      F(L+1,D)=F(L,I)-Y
      F(L+1,K)=F(L,K)+Z
      F(L+1,E)=F(L,K)-Z
      K=K-1
      END DO
      E=K+S1
      END DO
      S1=S2
      END DO

C  GET DFT
      R(Ø)=(F(L,Ø)+F(L,Ø))/2.
      X(Ø)=Ø
      DO I=1,N7
      B=F(L,N-I)
      R(I)=(F(L,I)+B)/2.
      X(I)=(F(L,I)-B)/2.
      END DO
4Ø    CONTINUE

C  WRITE RESULTS
      TYPE 5Ø
5Ø    FORMAT (' Input H( ) R( ) X( ) ')
      DO I=Ø,N7
      TYPE 6Ø, I,FIN(I),F(L,I)/N,R(I)/N,X(I)/N
6Ø    FORMAT (I6,4 F9.3)
      END DO
      END
```

Sample printout for $P = 8, N = 256$

```
.      Input     H( )      R( )      X( )
. 0    1.000   128.500   128.500     0.000
. 1    2.000   -41.242    -0.500   -40.742
. 2    3.000   -20.868    -0.500   -20.368
. 3    4.000   -14.075    -0.500   -13.575
. 4    5.000   -10.678    -0.500   -10.178
. 5    6.000    -8.639    -0.500    -8.139
. 6    7.000    -7.278    -0.500    -6.778
. 7    8.000    -6.306    -0.500    -5.806
. 8    9.000    -5.577    -0.500    -5.077
. 9   10.000    -5.009    -0.500    -4.509
.      .....    ......     .....     .....
```

Notes. *(a)* FORTRAN *version of* "FHTBAS". *(b) Delivers DHT as* F(P - 1,I) *and DFT as* R(I) + *i*X(I) *for* I $= 0$ *to* $N - 1$. *(c)*

```
C  FHTFOR.FOR
C  CALLING PROGRAM FOR FHTFOR
C
      COMMON NØ
      DIMENSION F(1024)
      P = 5

C  INSERT DATA
      DO I = 1,64
      F(I) = I
      END DO
      CALL FHTFOR(P,F)
      WRITE(5,2Ø)F
      STOP
      END

C  SUBROUTINE FHTFOR TAKES INPUT F() AND RETURNS THE DHT
C  IN THE SAME F()
C  LENGTH OF F() IS 2**P
      SUBROUTINE FHTFOR(P,F)
      COMMON NØ
      DIMENSION F(1Ø24)
9Ø2Ø   P1 = 3.141592
9Ø3Ø   IF (P.NEQ.1) GO TO 9Ø4Ø
9Ø33   J = F(Ø) + F(1)
9Ø34   F(1) = F(Ø) - F(1)
9Ø35   F(Ø) = J
9Ø36   RETURN
9Ø4Ø   CONTINUE
9Ø4Ø   DIMENSION S9(256), T9(256)
9Ø5Ø   INTEGER A9(64), M9(1Ø)
9Ø6Ø   DIMENSION V9(1Ø), C9(1Ø)
9Ø7Ø   N9 = 2**(P - 2)
9Ø71   N = 4*N9
9Ø72   C9(5) = N - 1
9Ø73   C9(6) = P - 1
9Ø9Ø   IF N(.EQ.)NØ GO TO 94ØØ

C  GET POWERS OF 2
92Ø1   I = 1
92Ø2   M9(Ø) = 1
92Ø3   M9(1) = 2
92Ø4   M9(I + 1) = M9(I) + M9(I)
92Ø5   I = I + 1
92Ø6   IF I(.LE.)P GO TO 92Ø4
C  SPECIAL CASE
9296   IF N(.EQ.)2 GO TO 9411
C  SKIP TRIGONOMETRIC FUNCTIONS
9297   IF N(.LE.)8 GO TO 94ØØ
9298   S9(N9) = 1
C  SKIP SINES
9299   IF (N.NEQ.8) GO TO 93ØØ
       S9(1) = SIN(P1/4)
```

```
      GO TO 933Ø

C  GET SINES
93ØØ   CONTINUE
C  COARSE SEED TABLE FOR SINES
93Ø1   DO I = 1,3
       S9(I*N9/4) = SIN(I*P1 /8)
       END DO
C  INITIAL HALF SECANT
93Ø2   H9 = 1/2/COS(P1/16)
C  FILL SINE TABLE
93Ø4   C9(4) = P - 4
93Ø5   DO I = 1,P - 4
93Ø6   C9(4) = C9(4) - 1
       V9(Ø) = Ø
93Ø7   DO J = M9(C9(4)),N9 - M9(C9(4)),M9(C9(4) + 1)
93Ø8   V9(1) = J + M9(C9(4))
93Ø9   S9(J) = H9*(S9(V9(1)) + V9(Ø))
       V9(Ø) = S9(V9(1))
931Ø   END DO
C  HALF SECANT RECURSION
9311   H9 = 1/SQRT(2 + 1/H9)
9312   END DO

C  GET TANGENTS
934Ø   C9(Ø) = N9 - 1
935Ø   DO I = 1,N9 - 1
936Ø   T9(I) = (1 - S9(C9(Ø)))/S9(I)
937Ø   C9(Ø) = C9(Ø) - 1
938Ø   END DO
9381   T9(N9) = 1
94ØØ   CONTINUE

C  FAST PERMUTE
C  FOR P = 2, 3 PERMUTE DIRECTLY
94Ø3   IF P(.NEQ.)2 GO TO 94Ø4
       V9(9) = F(1)
       F(1) = F(2)
       F(2) = V9(9)
       GO TO 95ØØ
94Ø4   CONTINUE
       IF P(.NEQ.)3 GO TO 94Ø5
       V9(9) = F(1)
       F(1) = F(4)
       F(4) = V9(9)
       V9(9) = F(3)
       F(3) = F(6)
       F(6) = V9(9)
94Ø5   IF P(.EQ.)3 GO TO 95ØØ
C  FOR P = 4, 5, 6 (Q9 = 2, 3), SKIP STRUCTURE TABLE
94Ø7   Q9 = P DIV 2
       C9(2) = M9(Q9)
94Ø8   Q9 = Q9 + MOD(P,2)
```

```
9409   IF Q9(.NEQ.)2 GO TO 9410
       A9(1) = 2
       A9(2) = 1
       A9(3) = 3
       GO TO 9420
9410   CONTINUE
       IF Q9(.NEQ.)3 GO TO 9411
       A9(1) = 4
       A9(2) = 2
       A9(3) = 6
       A9(4) = 1
       A9(5) = 5
       A9(6) = 3
       A9(7) = 7
       GO TO 9420
9411   CONTINUE
9411   IF N(.NEQ.)2 GO TO 9412
C  SPECIAL CASE
       V9(6) = F(0)
       F(0) = F(1)
       F(1) = V9(6)
9412   CONTINUE
C  SET UP STRUCTURE TABLE
9413   A9(0) = 0
       A9(1) = 1
9414   DO I = 2,Q9
9415   DO J = 0,M9(I - 1) - 1
9416   A9(J) = A9(J) + A9(J)
9417   A9(J + M9(I - 1)) = A9(J) + 1
9418   END DO
9419   END DO
9420   CONTINUE

C  PERMUTE
9422   DO I = 1,C9(2) - 1
9423   V9(4) = C9(2)*A9(I)
9424   V9(5) = 1
       V9(6) = V9(4)
9425   V9(7) = F(V9(5))
       F(V9(5)) = F(V9(6))
       F(V9(6)) = V9(7)
9426   DO J = 1,A9(I) - 1
9427   V9(5) = V9(5) + C9(2)
       V9(6) = V9(4) + A9(J)
9428   V9(7) = F(V9(5))
       F(V9(5)) = F(V9(6))
       F(V9(6)) = V9(7)
9429   END DO
9430   END DO

C  STAGES 1 & 2
C  GET TWO-ELEMENT DHTs
9502   DO I = 0,N - 2,2
9503   V9(6) = F(I) + F(I + 1)
       V9(7) = F(I) - F(I + 1)
```

```
9505   F(I) = V9(6)
       F(I + 1) = V9(7)
9507   END DO
9508   IF P = 1 RETURN
C  GET FOUR-ELEMENT DHTs
9510   DO I = 0,N - 4,4
9511   V9(6) = F(I) + F(I + 2)
       V9(7) = F(I + 1) + F(I + 3)
9513   V9(8) = F(I) - F(I + 2)
       V9(0) = F(I + 1) - F(I + 3)
9515   F(I) = V9(6)
       F(I + 1) = V9(7)
       F(I + 2) V9(8)
       F(I + 3) = V9(9)
9519   END DO
9520   IF P(.EQ.)2 RETURN

C  STAGES 3, 4, ...
9601   U9 = C9(6)
       S9 = 4
9603   DO L9 = 2,C9(6)
9604   V9(2) = S9 + S9
       U9 = U9 - 1
       V9(3) = M9(U9 - 1)
9607   DO Q9 = 0,C9(5),V9(2)
9608   I = Q9
       D9 = I + S9
9610   V9(6) = F(I) + F(D9)
       V9(7) = F(I) - F(D9)
9612   F(I) = V9(6)
       F(D9) = V9(7)
       K9 = D9 - 1
9615   DO J = V9(3),N9,V9(3)
9616   I = I + 1
       D9 = I + S9
       E9 = K9 + S9
9617   V9(9) = F(D9) + F(E9)*T9(J)
9618   X9 = F(E9) - V9(9)
9621   V9(6) = F(I) + Y9
       V9(7) = F(I) - Y9
       V9(8) = F(K9) - X9
       V9(9) = F(K9) + X9
9625   F(I) = V9(6)
       F(D9) = V9(7)
       F(K9) = V9(8)
       F(E9) = V9(9)
       K9 = K9 - 1
9630   END DO
9631   E9 = K9 + S9
9632   END DO
9633   S9 = V9(2)
9634   END DO
C  REMEMBER
9635   N0= N
```

```
9636    RETURN
        END
```

Notes: (a) This subroutine assumes that user will divide throughout by N subsequently. (b) Variable names and line numbers in this subroutine are mainly the same as those used in **FHTSUB** *(c) Copyright © 1985 The Board of Trustees of the Leland Stanford Junior University.*

APPENDIX 2

ATLAS OF HARTLEY TRANSFORMS

"Aprés cela prenant vn point a difcretion dans la courbe, comme C, ..., ie tire de ce point C la ligne CB parallele a GA, & pourceque CB & BA font deux quantités indeterminées & inconnuëes, ie les nomme l'vne y & l'autre x."

René Descartes, Discours de la Méthode, 1637, p. 321.

[First use of cartesian coordinates.]

In the following atlas, functions of time $f(t)$ are presented on the left with the corresponding Hartley transform $H(f)$ on the right. The direct and inverse transform relations are

$$H(f) = \int_{-\infty}^{\infty} f(t) \operatorname{cas} 2\pi ft \, dt$$

$$f(t) = \int_{-\infty}^{\infty} H(f) \operatorname{cas} 2\pi ft \, dt.$$

The following abbreviations are used.

The cas *function*

$$\operatorname{cas} t = \cos t + \sin t$$

Unit rectangle function

$$\Pi(t) = \begin{cases} 1, & |t| < \frac{1}{2}; \\ 0, & |t| > \frac{1}{2}. \end{cases}$$

Unit triangle function

$$\Lambda(t) = \begin{cases} 1 - |t|, & |t| < 1; \\ 0, & \text{elsewhere.} \end{cases}$$

Heaviside unit step function

$$\mathrm{H}(t) = \begin{cases} 1, & t > 0; \\ 0, & t < 0. \end{cases}$$

Ticks on the graphs refer to unit distance.

In addition to the examples for which analytic expressions are presented there are a number of waveforms in the form of polygons passing through the points (τ, f_τ). The values of the ordinates f_τ are integers or simple fractions and may be read off from the scale. Analytic expressions for the transforms are not given but may be obtained if desired from the summation

$$H(f) = \operatorname{sinc}^2 f \sum_{\tau} f_\tau \operatorname{cas} 2\pi f \tau.$$

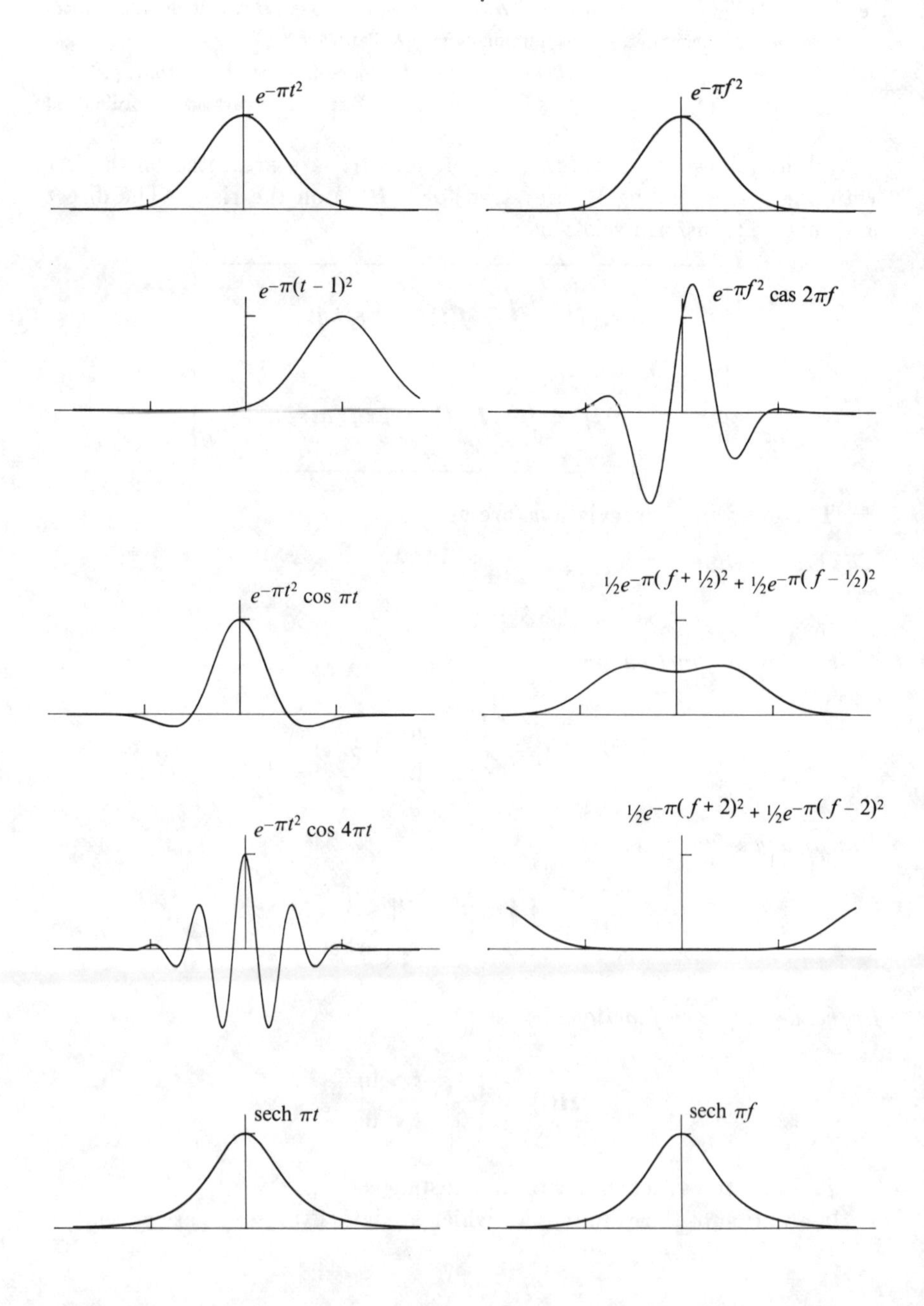

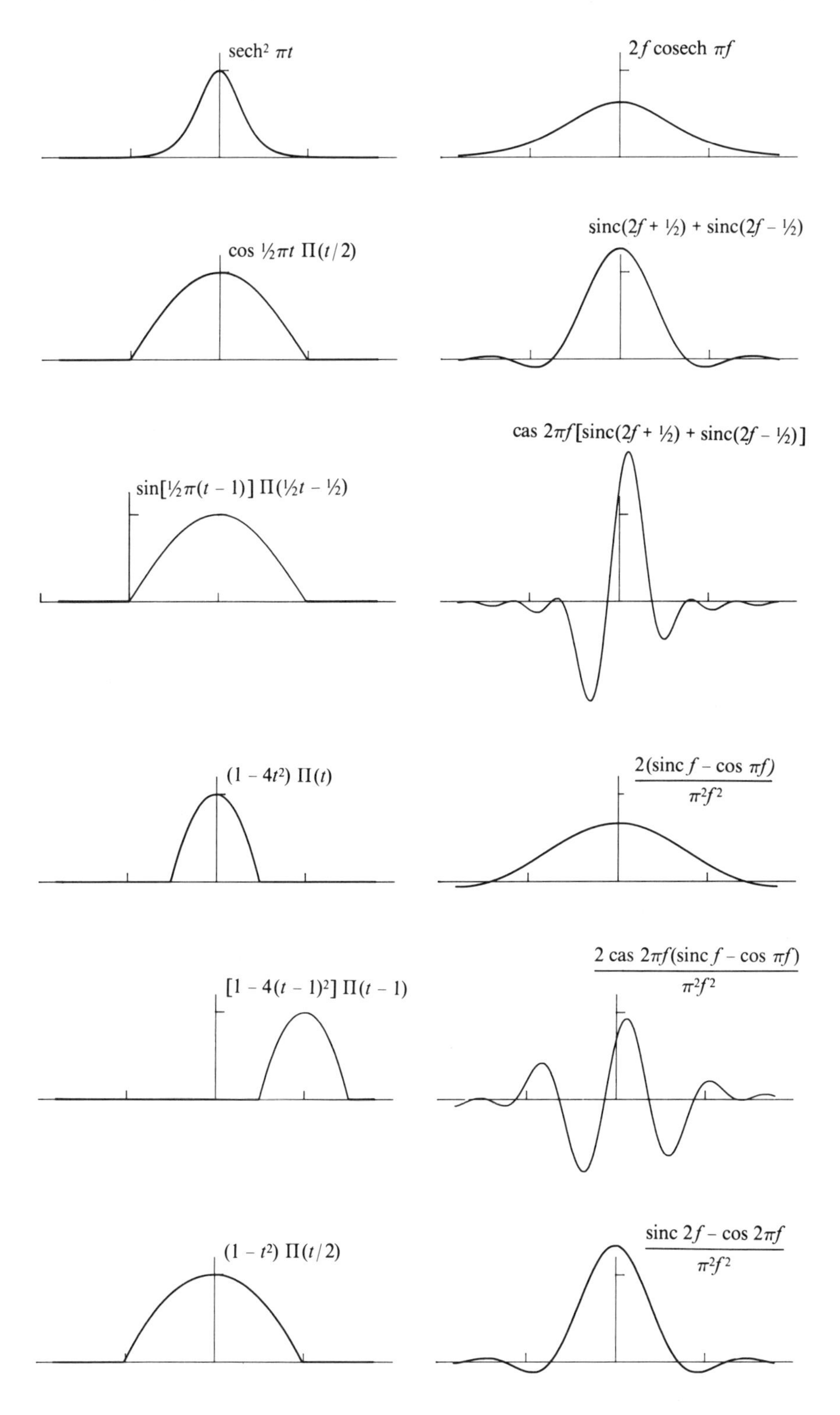

$\mathrm{sech}^2\, \pi t$
$2f\, \mathrm{cosech}\, \pi f$
$\cos \frac{1}{2}\pi t\, \Pi(t/2)$
$\mathrm{sinc}(2f + \frac{1}{2}) + \mathrm{sinc}(2f - \frac{1}{2})$
$\sin[\frac{1}{2}\pi(t-1)]\, \Pi(\frac{1}{2}t - \frac{1}{2})$
$\mathrm{cas}\, 2\pi f[\mathrm{sinc}(2f + \frac{1}{2}) + \mathrm{sinc}(2f - \frac{1}{2})]$
$(1 - 4t^2)\, \Pi(t)$
$\frac{2(\mathrm{sinc}\, f - \cos \pi f)}{\pi^2 f^2}$
$[1 - 4(t-1)^2]\, \Pi(t-1)$
$\frac{2\, \mathrm{cas}\, 2\pi f(\mathrm{sinc}\, f - \cos \pi f)}{\pi^2 f^2}$
$(1 - t^2)\, \Pi(t/2)$
$\frac{\mathrm{sinc}\, 2f - \cos 2\pi f}{\pi^2 f^2}$

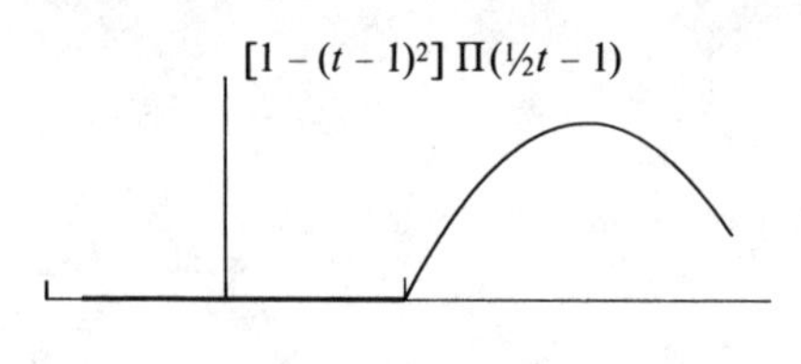
$[1 - (t - 1)^2]\,\Pi(\frac{1}{2}t - 1)$

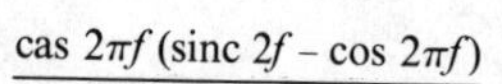
$\dfrac{\operatorname{cas} 2\pi f\,(\operatorname{sinc} 2f - \cos 2\pi f)}{\pi^2 f^2}$

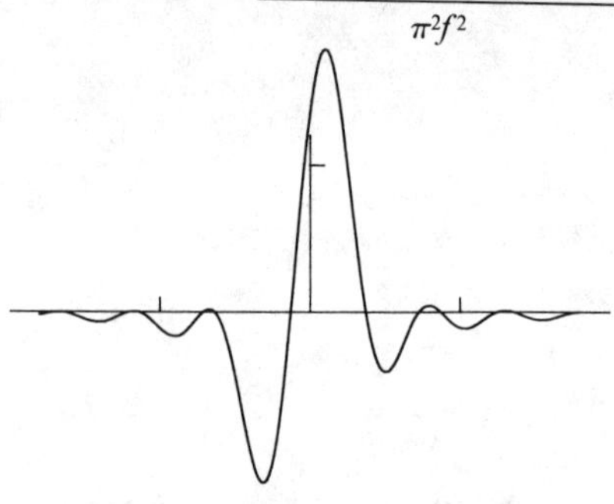

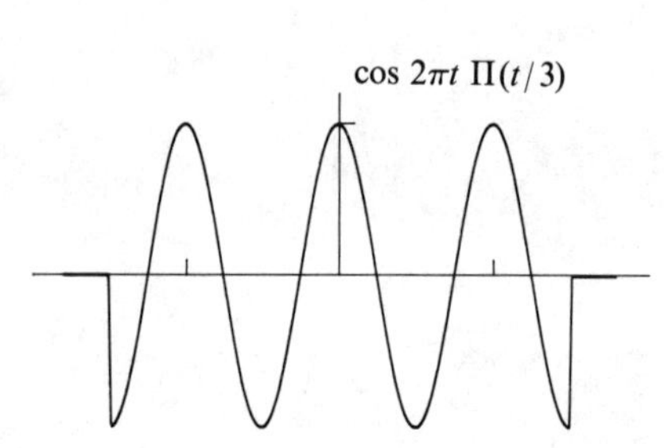
$\cos 2\pi t\ \Pi(t/3)$

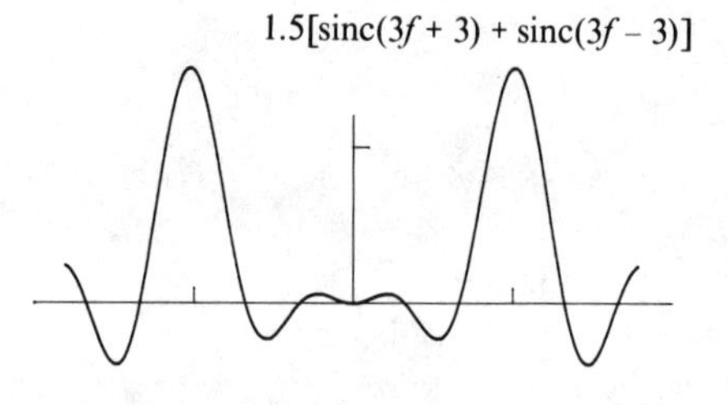
$1.5[\operatorname{sinc}(3f + 3) + \operatorname{sinc}(3f - 3)]$

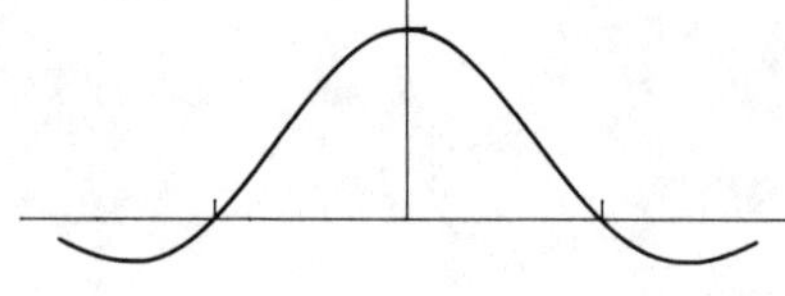
sinc t

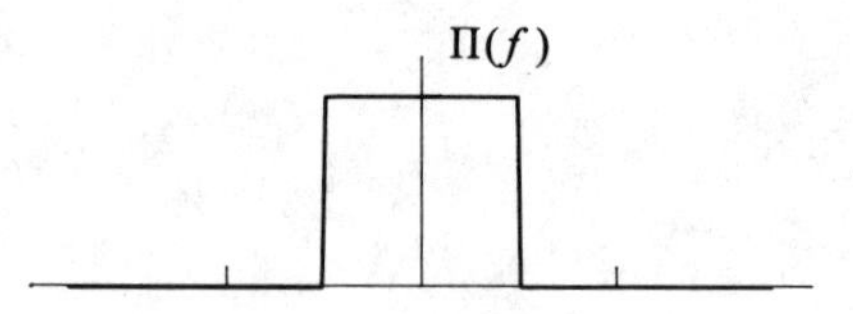
$\Pi(f)$

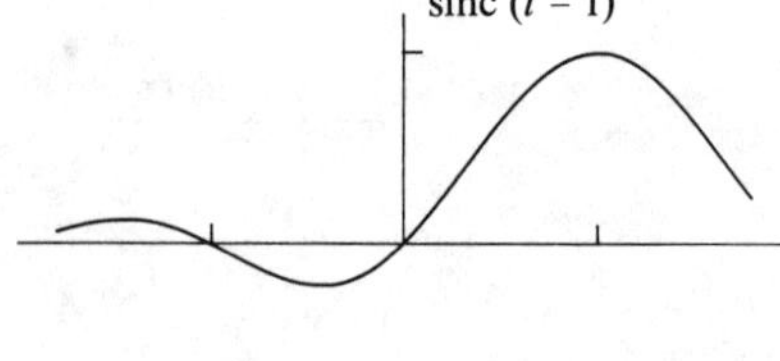
sinc $(t - 1)$

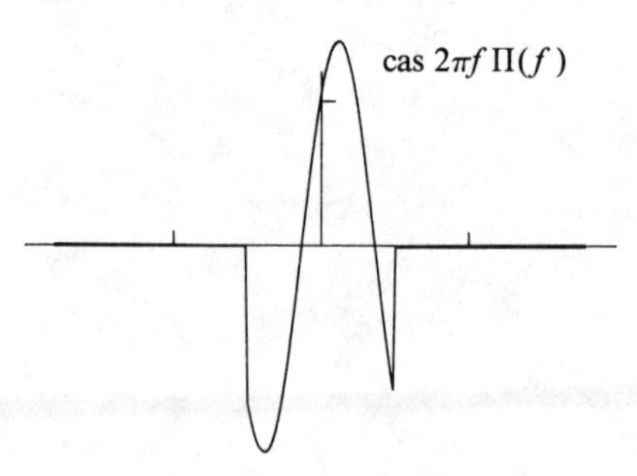
$\operatorname{cas} 2\pi f\ \Pi(f)$

$\Pi(t - \frac{1}{2})$

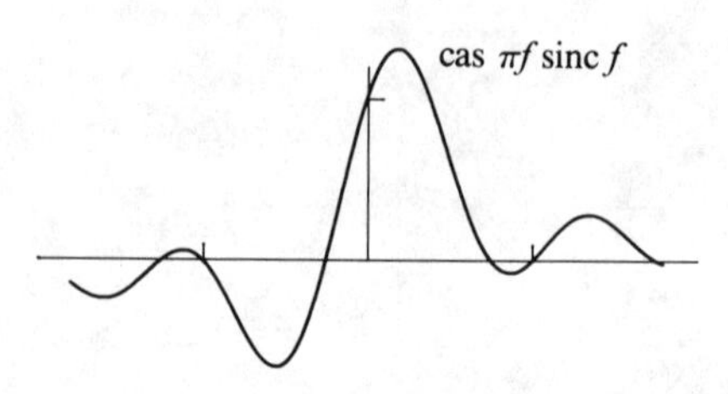
$\operatorname{cas} \pi f \operatorname{sinc} f$

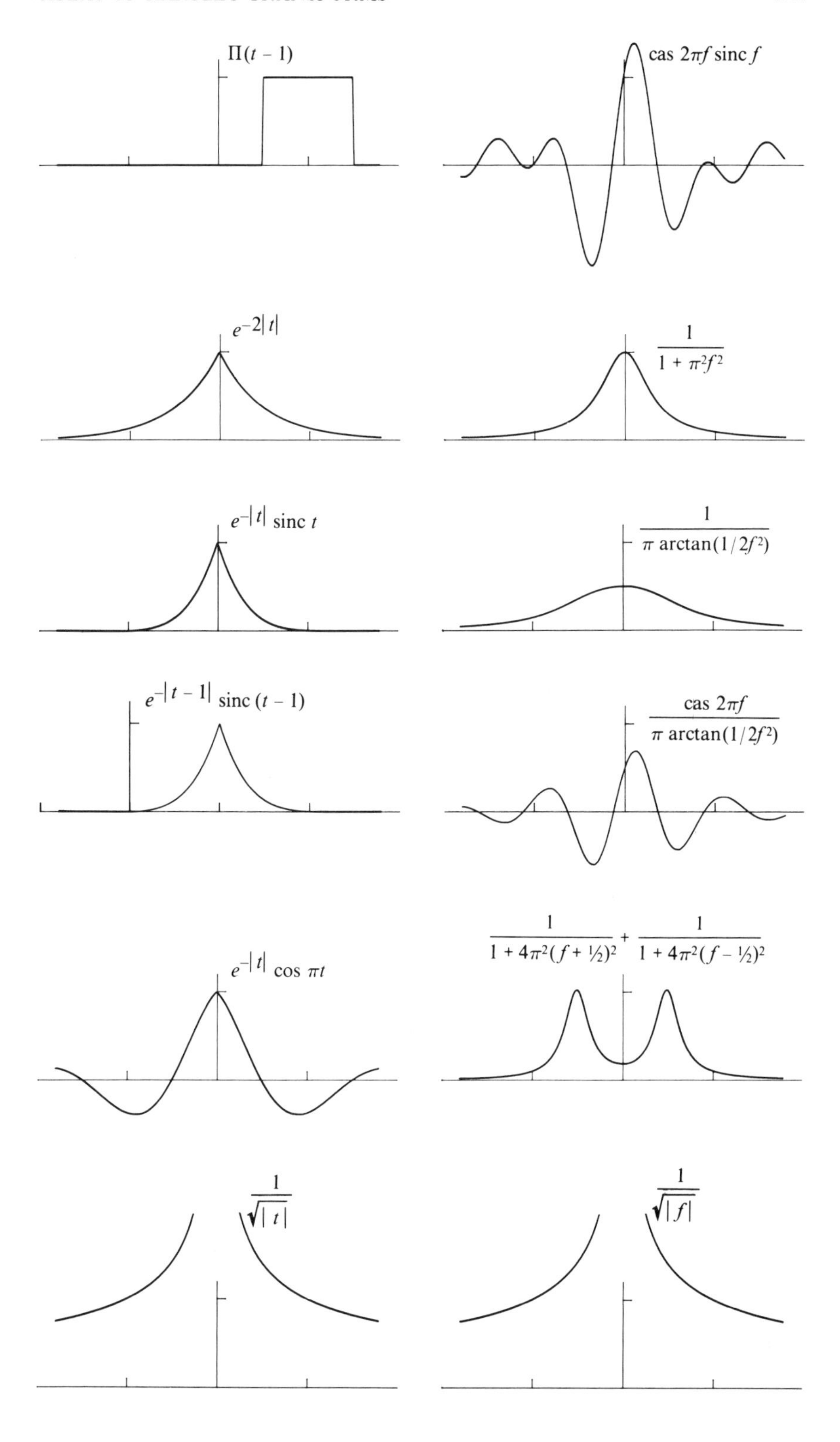

$\Pi(t-1)$
$\operatorname{cas} 2\pi f \operatorname{sinc} f$
$e^{-2|t|}$
$\frac{1}{1+\pi^2 f^2}$
$e^{-|t|} \operatorname{sinc} t$
$\frac{1}{\pi} \arctan(1/2f^2)$
$e^{-|t-1|} \operatorname{sinc}(t-1)$
$\frac{\operatorname{cas} 2\pi f}{\pi} \arctan(1/2f^2)$
$e^{-|t|} \cos \pi t$
$\frac{1}{1+4\pi^2(f+\frac{1}{2})^2} + \frac{1}{1+4\pi^2(f-\frac{1}{2})^2}$
$\frac{1}{\sqrt{|t|}}$
$\frac{1}{\sqrt{|f|}}$

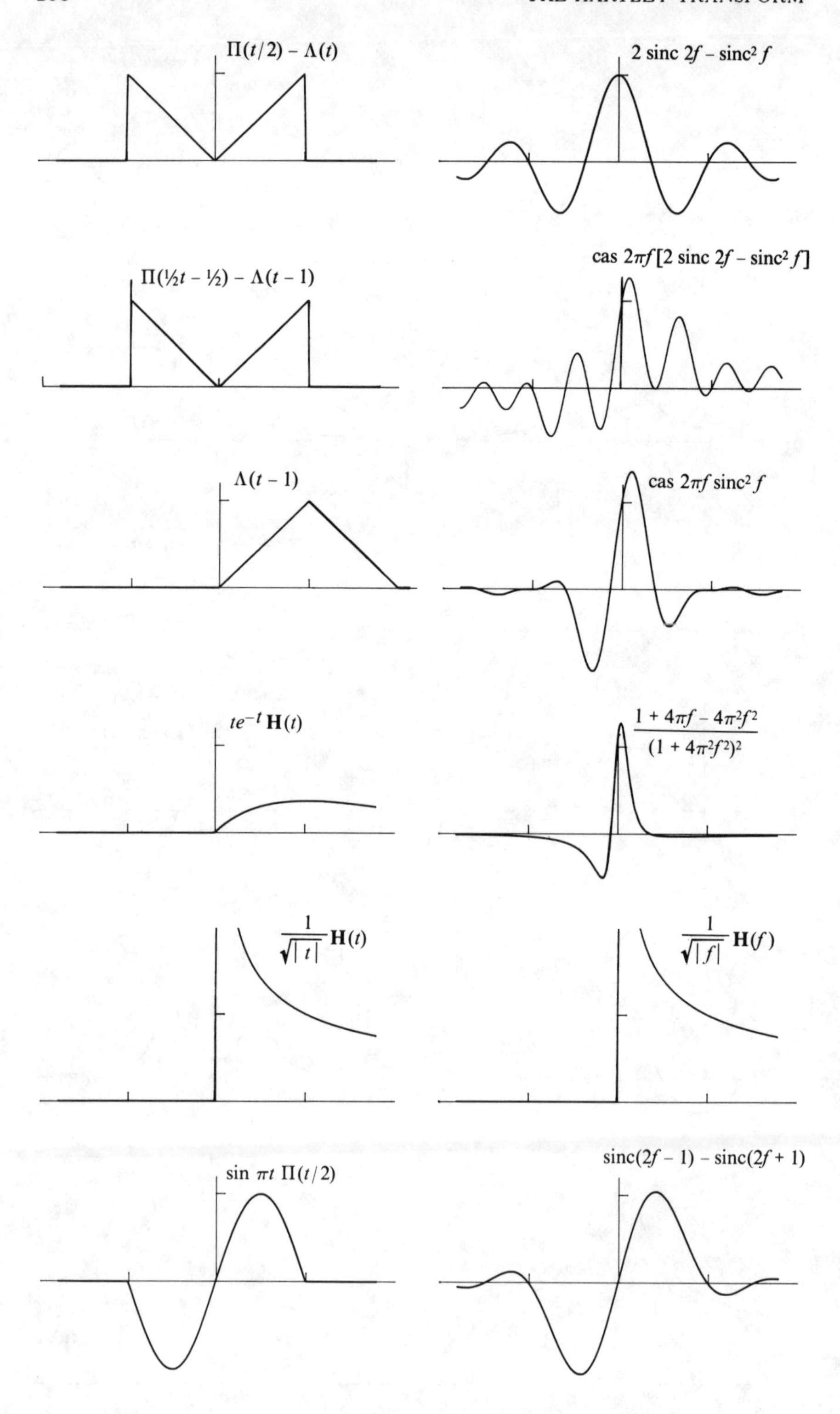
$\Pi(t/2) - \Lambda(t)$
$2 \operatorname{sinc} 2f - \operatorname{sinc}^2 f$
$\Pi(\frac{1}{2}t - \frac{1}{2}) - \Lambda(t-1)$
$\operatorname{cas} 2\pi f[2 \operatorname{sinc} 2f - \operatorname{sinc}^2 f]$
$\Lambda(t-1)$
$\operatorname{cas} 2\pi f \operatorname{sinc}^2 f$
$te^{-t}\mathbf{H}(t)$
$\frac{1 + 4\pi f - 4\pi^2 f^2}{(1 + 4\pi^2 f^2)^2}$
$\frac{1}{\sqrt{|t|}}\mathbf{H}(t)$
$\frac{1}{\sqrt{|f|}}\mathbf{H}(f)$
$\sin \pi t\, \Pi(t/2)$
$\operatorname{sinc}(2f-1) - \operatorname{sinc}(2f+1)$

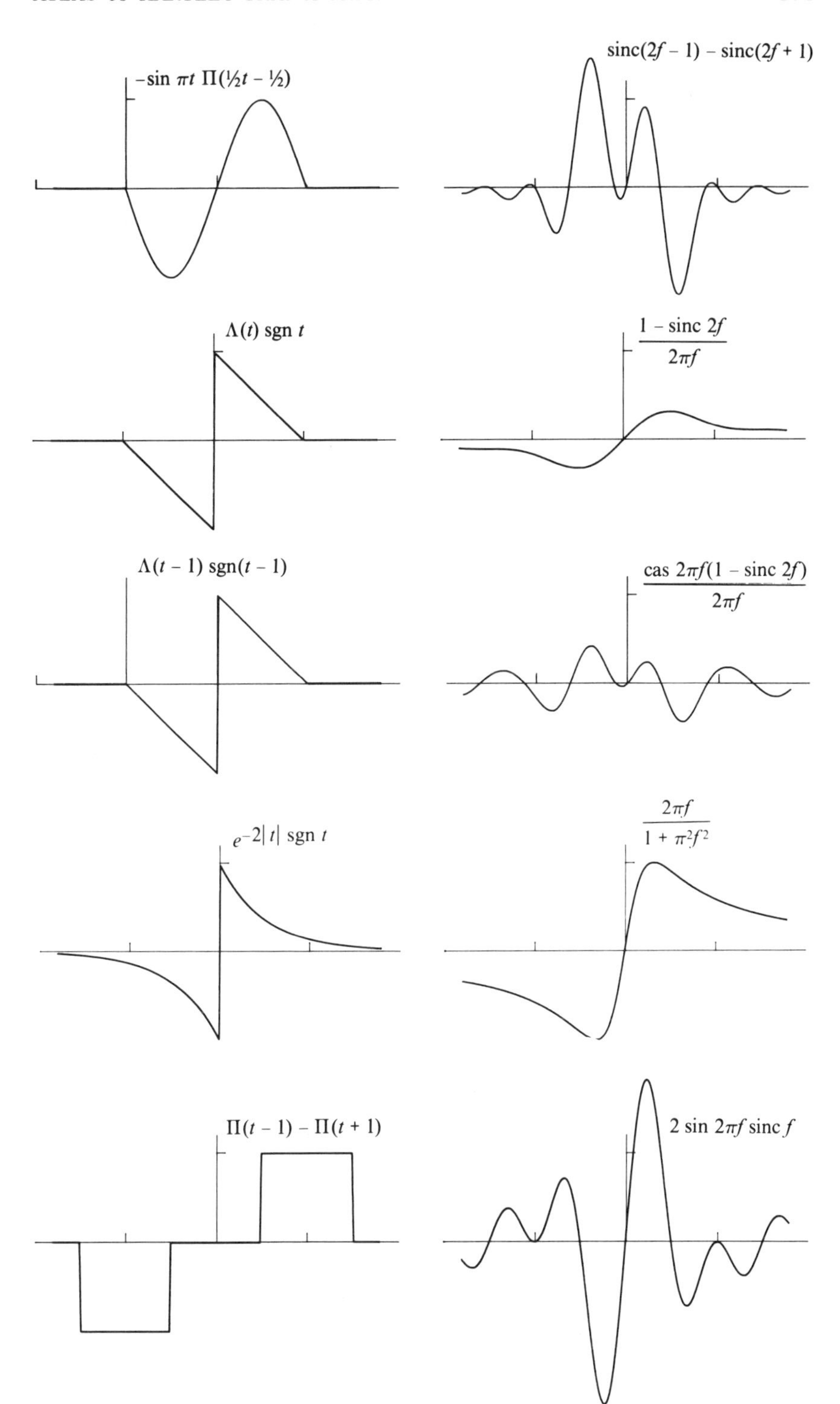

$-\sin \pi t\, \Pi(\frac{1}{2}t - \frac{1}{2})$
$\mathrm{sinc}(2f - 1) - \mathrm{sinc}(2f + 1)$
$\Lambda(t)\ \mathrm{sgn}\ t$
$\frac{1 - \mathrm{sinc}\ 2f}{2\pi f}$
$\Lambda(t - 1)\ \mathrm{sgn}(t - 1)$
$\frac{\mathrm{cas}\ 2\pi f(1 - \mathrm{sinc}\ 2f)}{2\pi f}$
$e^{-2|t|}\ \mathrm{sgn}\ t$
$\frac{2\pi f}{1 + \pi^2 f^2}$
$\Pi(t - 1) - \Pi(t + 1)$
$2 \sin 2\pi f\ \mathrm{sinc}\ f$

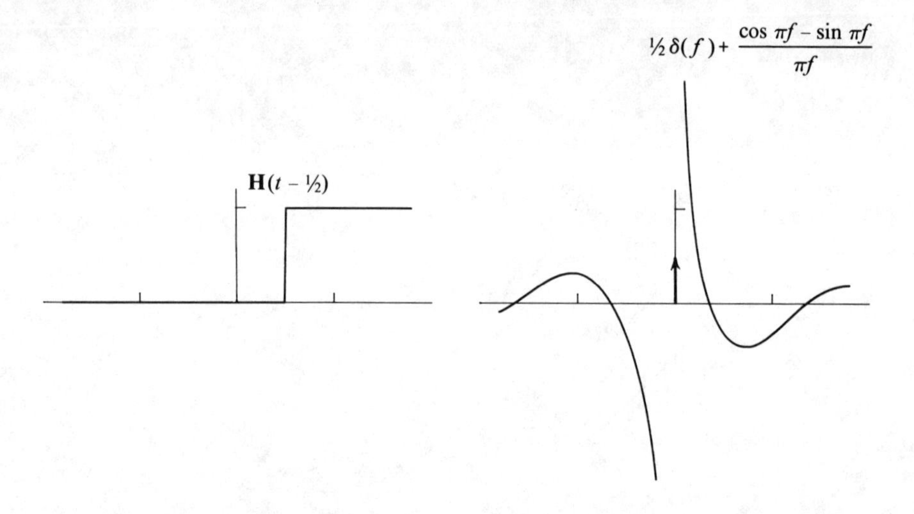
$\mathbf{H}(t - ½)$
$½\delta(f) + \frac{\cos \pi f - \sin \pi f}{\pi f}$

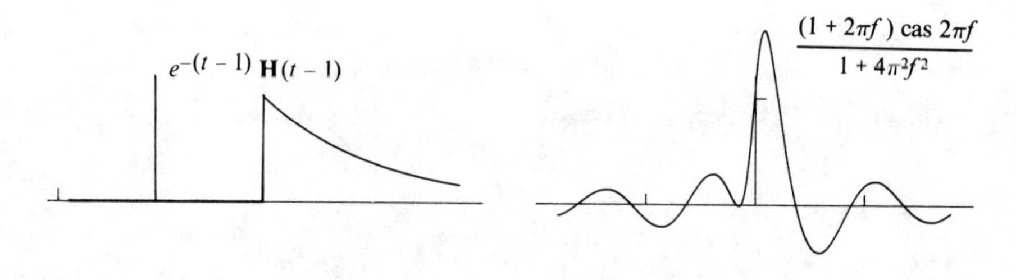
$e^{-(t-1)}\,\mathbf{H}(t - 1)$
$\frac{(1 + 2\pi f)\,\text{cas}\,2\pi f}{1 + 4\pi^2 f^2}$

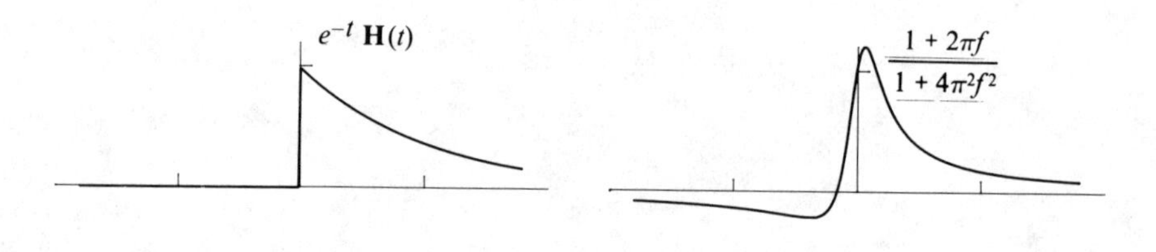
$e^{-t}\,\mathbf{H}(t)$
$\frac{1 + 2\pi f}{1 + 4\pi^2 f^2}$

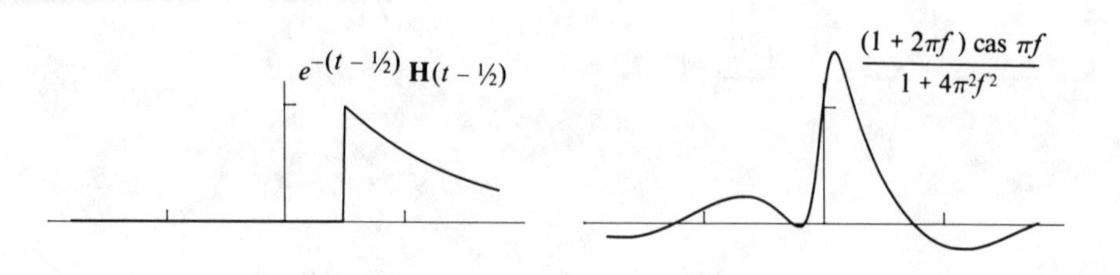
$e^{-(t-½)}\,\mathbf{H}(t - ½)$
$\frac{(1 + 2\pi f)\,\text{cas}\,\pi f}{1 + 4\pi^2 f^2}$

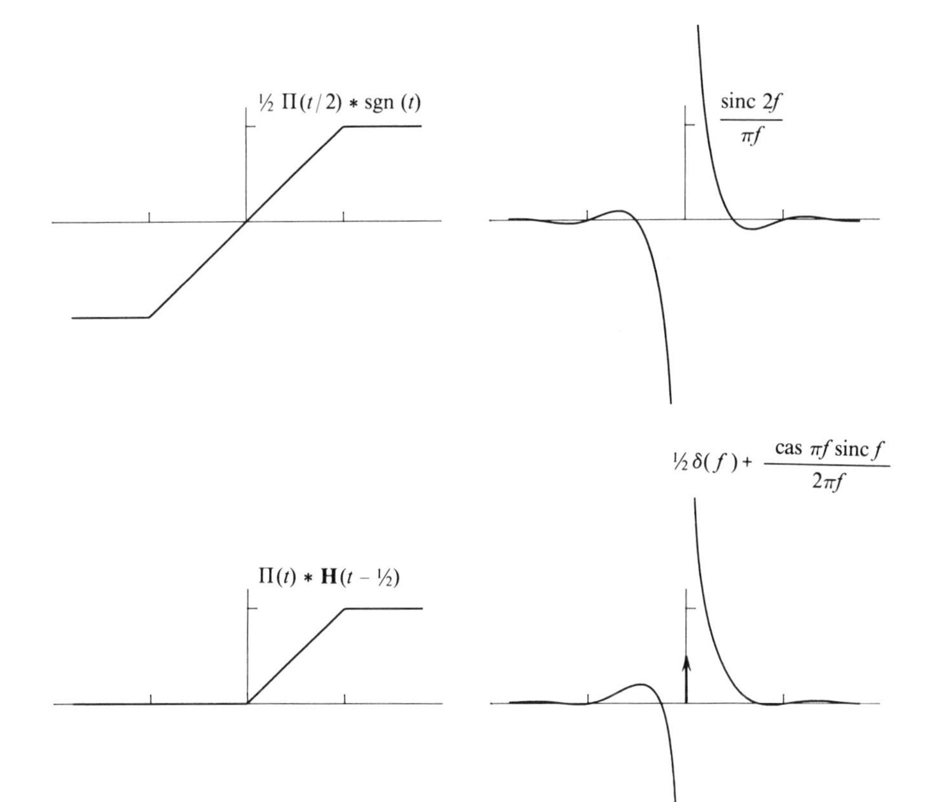

0 0 0 0 0 1 1 0 0 0 0 SUM=2

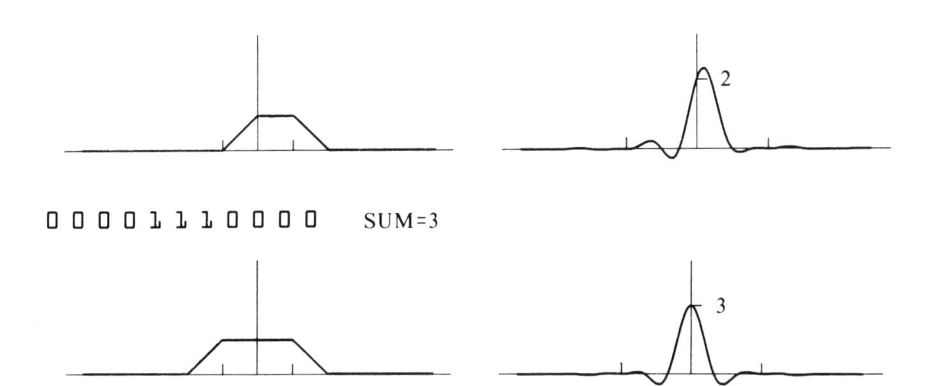

0 0 0 0 0 0 0 1 1 1 0 SUM=3

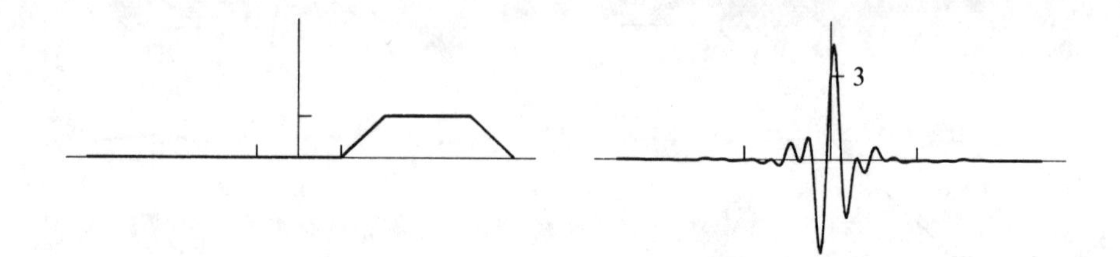

0 0 0 0 0 1 1 1 1 0 0 SUM=4

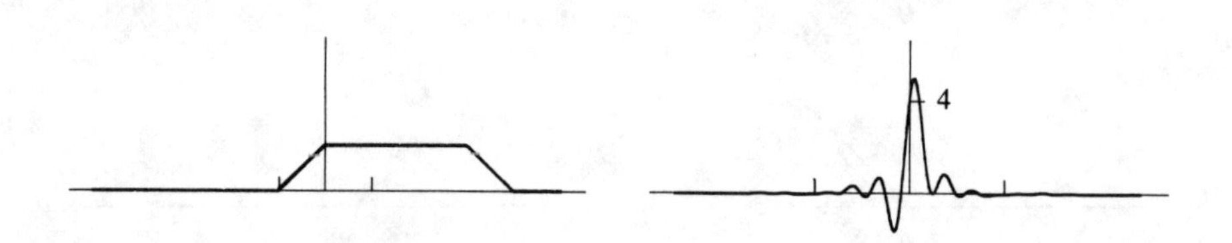

0 0 0 2.5 2 1.5 1 0.5 0 0 0 SUM=7.5

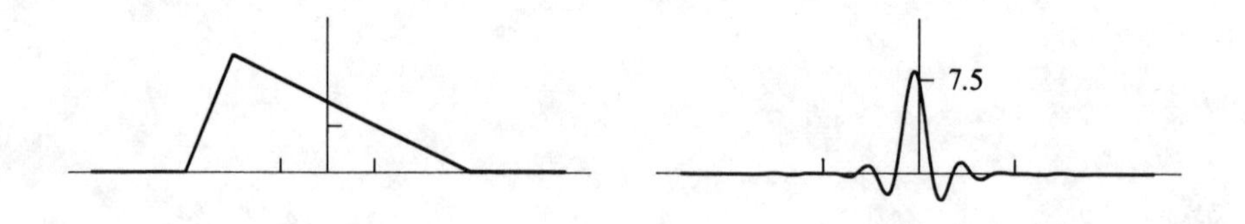

0 0 0 0 0 1 1 1 0 0 0 SUM=3

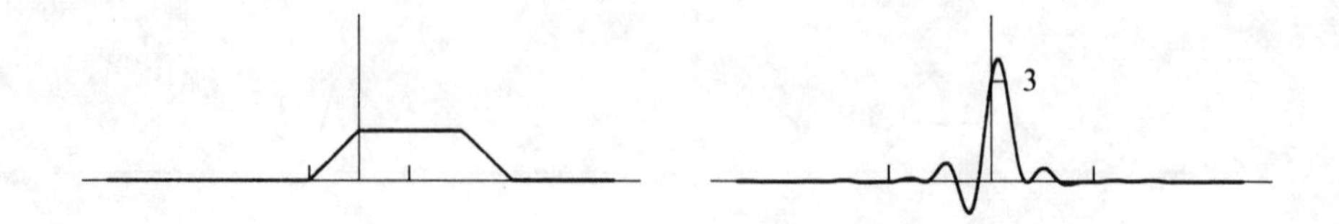

0 0 0 0 0 0 1 1 1 0 0 SUM=3

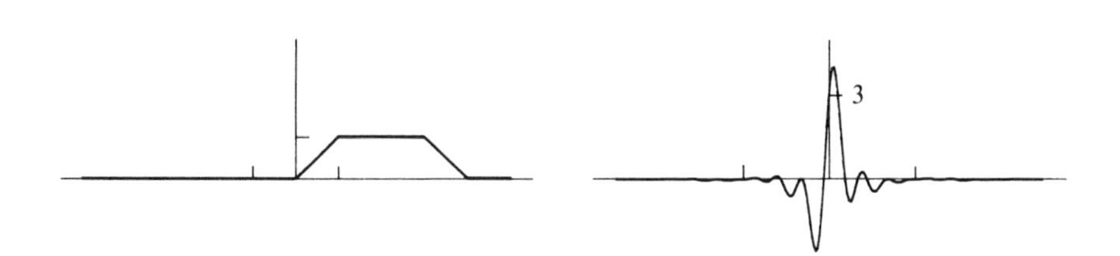

0 0 0 0 0 1 2 1 0.5 0.25 0 SUM=4.75

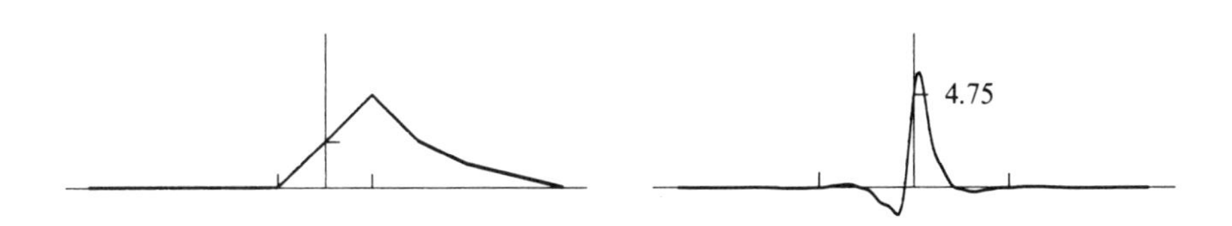

0 0 2 -0.999 0.5 -0.249 0.125 -0.0 SUM=1.328125

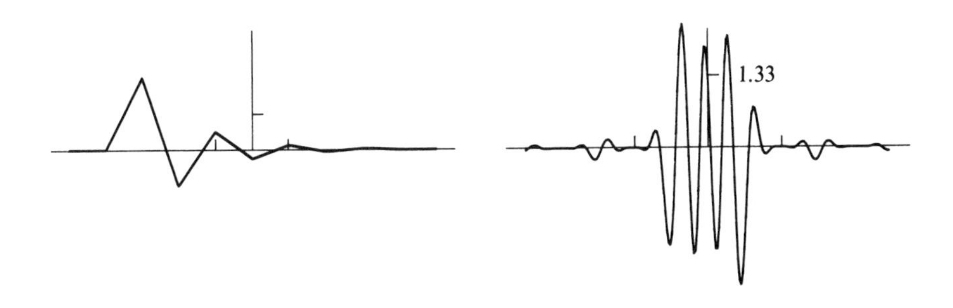

INDEX